Magnetic Communications

The ideal reference book providing all the information needed to fully understand magnetic communications in a self-contained source, written by experts in the field. This book offers a comprehensive introduction to magnetic communication using easy-to-understand language to explain concepts throughout and introduces the theory step by step with examples. A careful balance of combined theoretical and practical perspective is given throughout the book with interdisciplinary and multidisciplinary considerations for an in-depth and diverse understanding. This book covers the background, developments, fundaments, antennas, channels, performance, protocol related to magnetic communications, as well as applications that are of current interest, such as Internet of Things (IoT), multiple-input multiple-output (MIMO), and wireless power transfer. The figures of merit within magnetic communication system components are included, demonstrating how to both model and analyze them. This book will be of great benefit to graduate students, researchers, and electrical engineers working in the fields of wireless communications and the IoT.

Erwu Liu is an IET (Institute of Engineering and Technology) fellow and the founding Editor-in-Chief of IET/Wiley Blockchain. His team developed the first magnetic communication system in China and has won various awards, including the Special Award in the China International Industry Fair (2014) and the First Place in the Microsoft Indoor Localization Competition (2016).

Zhi Sun received his Ph.D. degree in Electrical and Computer Engineering from Georgia Institute of Technology (2011). He is a tenured associate professor at Tsinghua University. Prior to that, he was an associate professor at the University at Buffalo. Zhi Sun is a recipient of the US National Science Foundation (NSF) CAREER Award and a senior member of the Institute of Electrical and Electronics Engineers (IEEE).

Rui Wang received his Ph.D. degree in Engineering from Shanghai Jiao Tong University. He is an associate professor in the Department of Computer Science and Engineering at Tongji University. Prior to that, he was with The Chinese University of Hong Kong, as a post-doctoral research associate between 2013 and 2014.

Hongzhi Guo received his Ph.D. degree in Electrical and Electronics Engineering from The State University of New York at Buffalo (2017). He is an assistant professor at Norfolk State University. Previously, he was an assistant professor at University of Southern Maine. Hongzhi Guo is a recipient of the US NSF CAREER Award and a senior member of IEEE.

Magnetic Communications

Theory and Techniques

ERWU LIU
Tongji University

ZHI SUN
Tsinghua University

RUI WANG
Tongji University

HONGZHI GUO
Norfolk State University

CAMBRIDGE
UNIVERSITY PRESS

CAMBRIDGE
UNIVERSITY PRESS

Shaftesbury Road, Cambridge CB2 8EA, United Kingdom

One Liberty Plaza, 20th Floor, New York, NY 10006, USA

477 Williamstown Road, Port Melbourne, VIC 3207, Australia

314–321, 3rd Floor, Plot 3, Splendor Forum, Jasola District Centre, New Delhi – 110025, India

103 Penang Road, #05–06/07, Visioncrest Commercial, Singapore 238467

Cambridge University Press is part of Cambridge University Press & Assessment, a department of the University of Cambridge.

We share the University's mission to contribute to society through the pursuit of education, learning and research at the highest international levels of excellence.

www.cambridge.org
Information on this title: www.cambridge.org/9781108481670

DOI: 10.1017/9781108674843

First published 2024

A catalogue record for this publication is available from the British Library

A Cataloging-in-Publication data record for this book is available from the Library of Congress

ISBN 978-1-108-48167-0 Hardback

Contents

Preface

Using the near field of a coil to transmit a signal instead of propagation waves, magnetic communication turns out to be an effective and reliable communication approach for extremely harsh environments that are hostile to wireless communications such as sensor networking underground or underwater. Magnetic communication technologies have not only the great significance for the development of communication theory but also the great potential for a large number of engineering applications in the Internet of Things (IoT) era.

The so-called magnetic communication makes use of the time-varying magnetic field produced by the transmitting antenna, so that the receiving antenna receives the energy signal by mutual inductance. Researches show that the penetrability of a magnetic communication system depends on the magnetic permeability of the medium. Because the magnetic permeability of the layer, rock, ice, soil, and ore bed is close to the air, channel conditions bring less effects to magnetic transmission than electric transmission. Therefore, the communication network based on deep-penetrating magnetic induction (MI) can expand the perception ability and sensing range of information technology effectively, which can be applied to complex environments, such as underground, underwater, tunnel, mountain, rock, ice, and forest. The subject is currently receiving great scientific attention, and several important new results have been recently obtained. This book will present a comprehensive account of this emerging field.

The book comprises three parts.

- Part I: Introduction and Properties of MI
  Two initial chapters (Chapters 1–2) of this part give an introduction about magnetic communications. Chapter 1 presents the history, state of the art, and research challenges of magnetic communications. Chapter 2 briefly discusses the differences between magnetic communications and other wireless communications. It also describes some basic performance indicators of magnetic communications. This chapter constitutes the reference point for appreciating the results that MI holds.
  Subsequently, two chapters are devoted to antenna technology for magnetic communication (Chapter 3) and channel characteristics (Chapter 4).

- Part II: Theoretical Basis
  This part mainly expounds the theory of magnetic communication from different aspects. Specifically, the MI connection (Chapter 5), network performance (Chapter 6), and protocol stacks (Chapter 7) are analyzed.

- Part III: Applications
  This part (Chapter 8) discusses certain MI applications to validate the previously introduced theories.

Recent years have seen a growing trend in the application of communication technologies in extremely harsh environments. Magnetic communication is usually necessary when we want reliable signal propagation through radio-frequency-hostile (RF-hostile) media. In principle, magnetic communication is complementary to wireless communications. This is why it is important to understand the figures of merit of the components of a magnetic communication system, how to model them, and how to analyze them. Moreover, with the fast penetration of the IoT services into RF-hostile environments, an in-depth understanding of magnetic communication technologies is highly desired. Although several research papers on magnetic communications are available to date, the available information is very limited. It is the appropriate time to have a reference book providing all the information needed to understand magnetic communications in a single volume. We recognize that a single-volume book cannot cover all techniques, nor is it our intention to cover everything in one single book. Instead, we hope this book will serve our purpose of offering a first course on this important and booming subject of magnetic communications and will raise the interest needed to take magnetic communication research, development, and standardization activities to the next level.

This book will be of interest for graduate students, researchers, and electrical engineers working in the fields of wireless and the IoT. It will provide a comprehensive introduction to magnetic communication from a combined theoretical and practical perspective, and the readers are expected to have basic knowledge in wireless communications and electromagnetic theory.

Part I

Introduction and Properties of MI

1 Introduction

As one of the most vibrant areas in the communication field, wireless communication has been going through rapid development. On the other hand, there has been an explosive increase in demand for tetherless access in more and more scenarios. The Internet of Things (IoT), for instance, is an important development stage of information age. The vision of IoT is to achieve "Internet of Everything"; the key to the realization of the goal is to build a fully connected multiuser network that can be applied to diversified communications. As IoT applications are extending to more and more scenarios, common wireless communication techniques using electromagnetic (EM) waves are becoming dissatisfactory in both coverage and connectivity. EM waves experience high levels of attenuation due to absorption by a natural medium, such as soil, rock, and water, which leads to its inability to transmit in some challenging environments (underground, deep mine, mountain, rock, ice, tunnel, underwater, forest, . . .) as well as restrictions to the development of IoT.

To overcome this difficulty, deep penetration techniques, such as magnetic communications (MC), have brought solutions to these transmission problems. The so-called MC makes use of the time-varying magnetic field produced by the transmitting antenna, so that the receiving antenna receives the energy signal by mutual inductance. Researches show that the penetrability of the MC system depends on the magnetic permeability of the medium. Because the magnetic permeability of the layer, rock, ice, soil, and ore bed is close to the air, channel conditions bring less effects to magnetic transmission than electric transmission. Therefore, the communication network based on deep-penetrating magnetic induction (MI) can expand the perception ability and sensing range of information technology effectively, which can be applied to complex environments, such as underground, underwater, tunnel, mountain, rock, ice, and forest. We can conclude that the network construction of IoT based on MC is of great value and can be regarded as one of the reliable technologies to improve the connectivity of a wireless network.

The deep-penetrating MC technology is based on the principle of mutual inductance of the magnetic field. The alternating magnetic field is generated by the transmitting coil, and the receiving terminal also uses the coil antenna to measure the mutual inductance of the time-varying magnetic field in the space to obtain the information encoded in the time-varying signal (Fig. 1.1). In MC technology, in order to achieve reliable long-distance penetrating transmission, the sensitivity of the receiving antenna is usually required to reach the pT (pico Tesla) level. In addition, the performance

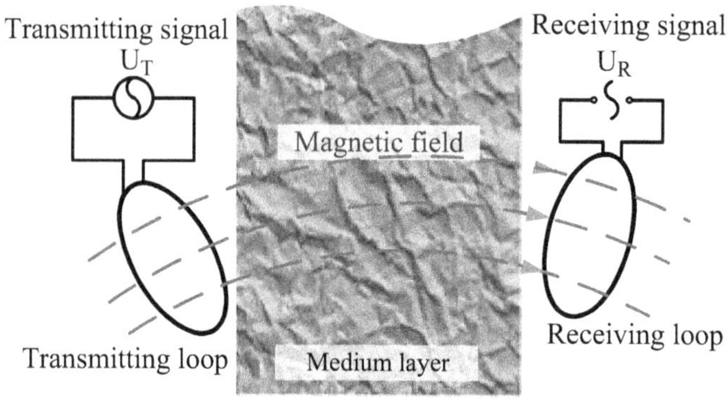

Figure 1.1 Wireless magnetic communication

optimization of the transmitter circuit in MC, antenna design, and receiving signal noise filtering are very challenging technical aspects.

Fundamental EM field and circuit theories for MC are included in this chapter. Advanced MC theory and latest MI applications are discussed in the following chapters.

1.1 Magnetic Communication and Wireless Communication

In the last two decades, driven by a wealth of theoretical development and practical requirement, many kinds of communication technologies under different situations drew the attention of research community. At present, the mainstream communication technologies include acoustic communication, optical waves communication, EM wave communication, and MC. This chapter covers the subject of the differences between MC and wireless communication.

1.1.1 The Comparison between Magnetic Communication and Other Wireless Communications

For most challenging environments mentioned in the beginning, there are no wireless communications deployment attempts before IoT applications. On the other hand, other wireless communication technologies are not able to provide reliable coverage and connectivity in such environments. Take underground mine for example, EM-wave-based wireless communication can only support a semi-wireless system, in which the links between the surface and underground are wired. Such a system is vulnerable, especially in a disaster situation.

However, in an underwater environment, many wireless communication technologies have been studied for both industry and military demands. The majority of the work on underwater communication is mainly based on acoustic communication, while it exhibits high propagation delays along with very low data rates, and highly

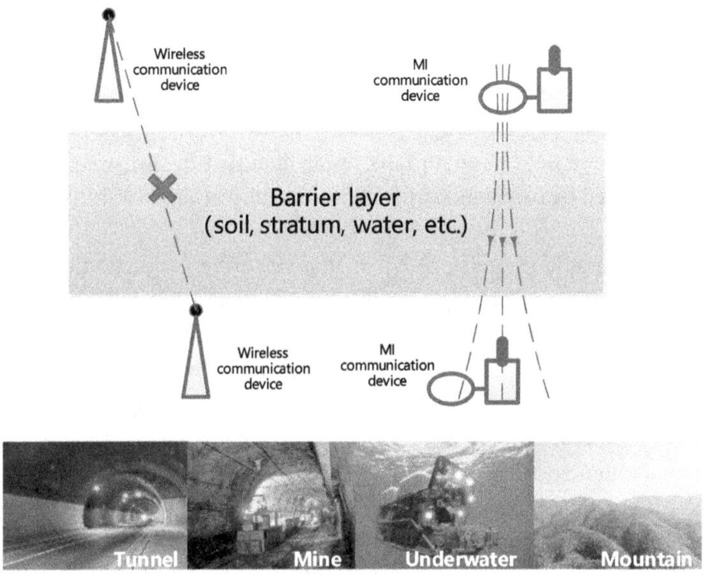

Figure 1.2 Communication scenarios under rugged environment

environment-dependent channel behavior. The highly environment-dependent channel behavior in underwater communication is caused by complex multipath fading, prevalent Doppler effects, and significant variation of these properties due to temperature, salinity, or pressure [1]. Optical waves do not suffer from high attenuation but experience multiple scattering of light, which results in inter-symbol interference and short transmission range [2]. Moreover, the transmission of optical signals requires high precision in pointing the narrow laser beams. Traditional wireless communication techniques using EM waves encounter three major problems: high path loss, dynamic channel condition, and large antenna size [3]. EM waves experience high attenuation that severely limits the achievable communication range. To increase the communication range, large antennas are required for low-frequency EM communication, which is not practical for small underwater vehicles and robots.

To sum up, the penetration ability of the traditional approaches is relatively weak, leading to propagation difficulties in a challenging environment, such as underground, deep mine, mountain area, terrane, tunnel, underwater, and forests, as shown in Fig. 1.2.

1.1.2 Benefits of Magnetic Communication

Magnetic communication is a promising alternative technique providing solutions for the mentioned problems. It utilizes the transmitting antennas to generate a time-varying magnetic fields in the medium, thus enabling the receiving antennas to receive the energy signal in a sense of mutual inductance. Our research shows that the dielectric penetration performance of the MC system depends mainly on the magnetic conductivity of the medium.

Using MI could have several benefits. One of these benefits is that dense media (such as soil and water) cause little variation in the rate of attenuation of magnetic fields from that of air, since the magnetic penetrability for each of these materials is almost the same [4]. Although generally unfavorable for open-air communication since the magnetic field strength falls faster than that in EM waves, the reduction in signal loss caused by propagation through soil compensates for this in the underground scenario.

Another favorable property of MI is that since the magnetic field is generated in the near field, it is non-propagating [4], which means that multipath fading is not a problem for MC. Moreover, since communication is achieved by coupling in the non-propagating near field, a transmitting device can detect the presence of any active receivers via the induced load on the coil. This property may provide valuable information for protocols, acting as a type of acknowledgement that the transmission was sensed by a remote device.

In an MC system, the antenna design is accomplished with the use of a coil of wire for both transmission and reception. The strength of the magnetic field produced by a given coil is proportional to the number of turns of wire, the cross-sectional area of the coil, and the magnetic permeability of any material placed in the core of the coil. The use of wire coils for MI transmission and reception represents a substantial benefit over the use of antennas for propagating EM waves. Low frequencies necessary for the propagation of EM waves mean that large antennas are necessary for reasonable efficiency, which obviously conflicts with the necessity that underground sensors remain small.

We take the underwater scenario, for instance, as shown in Table 1.1. Although the bandwidth of the MI system is smaller than that of the EM wave system, MC provides a longer transmission range. MC also has the advantage that its performance is not influenced by the properties of the medium.

Based on the advantages discussed above, an MC network can effectively expand the awareness and perception of the network. It has a good performance even in some harsh scenarios with many natural mediums or medium boundaries, such as underground, underwater, tunnels, massif, rock stratum, ice layer, and forest. The MI system enables a reliable and stable communication in some challenging environments instead of the EM system.

1.1.3 Applications of Magnetic Communication

Different from the traditional EM communication systems, the transmitting antenna of an MC system is equivalent to a magnetic dipole, which almost does not generate an electric field. Hence, the MI carrier is not a propagating wave, and it can be regarded as a quasi-static magnetic field generated in the air. So, the MI signals are free from the influence of multipath propagation compared with ordinary wireless signals. Due to the fact that the permeability of soil and water is close to that of air, MI signals

Table 1.1 Comparison of underwater MI, EM, acoustic, and optical communications

Communication paradigm	Propagation speed	Data rates	Communication ranges
MI	3.33×10^8 m/s	Mb/s	10–100 m
EM	3.33×10^8 m/s	Mb/s	≤10 m
Acoustic	1,500 m/s	kb/s	km
Optical	3.33×10^8 m/s	Mb/s	

Communication paradigm	Channel dependency	Stealth operation
MI	Conductivity	Yes
EM	Conductivity, multipath	Yes
Acoustic	Multipath, Dropper, temperature, pressure, salinity, environmental sound noise	Audible
Optical	Light scattering, line-of-sight communication, ambient light noise	Visible

can easily penetrate mediums such as water, sediment layer, and rock. Therefore, MC enable many important applications.

In [5], the authors introduced MI technologies to a wireless sensor network for underground pipeline monitoring. This MC system can provide a low-cost and real-time leakage detection and localization technique for underground pipelines. The authors of [6] and [7] analyzed the performance of an MC system underwater to measure basic communication metrics, such as the signal-to-noise ratio, bit error rate (BER), connectivity, and communication bandwidth. An MC system can also be applied to address the issue of water shortage confronting irrigation, which was studied in [8]. The authors of [8] used the MC network for Wireless Underground Sensor Networks (WUSN) instead of the EM wave communication for WUSN to realize an irrigation control system in horticulture in Australia. In a district heating system, MC technologies also play a big role in coping with the challenging underground channel environment discussed in [9]. In addition, MC technologies can be of great benefits in rescuing people if there's a mining disaster, flooding, or a collapse of underground tunnels [10].

Besides the automation and communication applications presented above, another important application of MC technologies is localization. MI localization does not rely on a propagating wave but generates a quasi-static magnetic field in the air. This direction has drawn much attention recently. A team at the University of Oxford developed an MI-based localization system that is shown in Fig. 1.3 to provide 3D localization in [11]. This incrementally deployed system can quickly localize a challenging underground scenario with accuracy around 1 m. In [12], the MI system was applied to indoor localization. MI localization has a huge advantage that obstacles, such as walls, floors, and people, which heavily impact the performance of EM waves are almost "transparent" to the MC system. However, the MI system has its own drawback, i.e., it

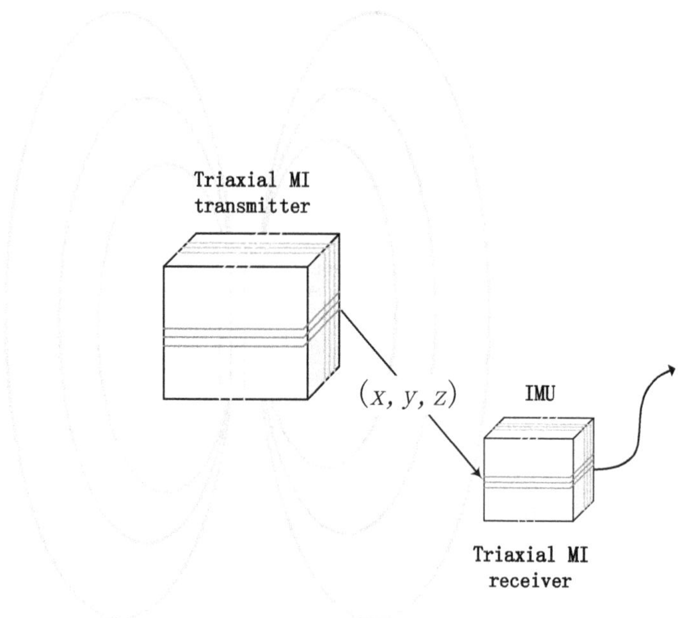

Figure 1.3 MI-based localization system

is sensitive to materials. By using signal processing and sensor fusion across multiple-system layers against the sensitivity to materials, the MI localization system can get 3D positioning with localization errors below 0.8 m even in some heavily distorted areas.

1.2 Preliminaries

1.2.1 Polar Coordinate

Polar coordinate has been frequently used in this book. As shown in Fig. 1.4, we use $\widehat{e}_r$, $\widehat{e}_\theta$, and $\widehat{e}_\phi$ to represent three unit coordinate vectors of a polar coordinate system. Let $\widehat{e}_x$, $\widehat{e}_y$, and $\widehat{e}_z$ be the unit coordinate vectors of a rectangular coordinate system. Then we have the following relationships:

$$
\begin{aligned}
\widehat{e}_r \cdot \widehat{e}_z &= \cos\theta \\
\widehat{e}_\theta \cdot \widehat{e}_z &= \cos\theta + \frac{\pi}{2} \\
\widehat{e}_\phi \cdot \widehat{e}_x &= \cos\phi + \frac{\pi}{2} \\
\widehat{e}_\phi \cdot \widehat{e}_y &= \cos\frac{\pi}{2} - \phi.
\end{aligned}
\tag{1.1}
$$

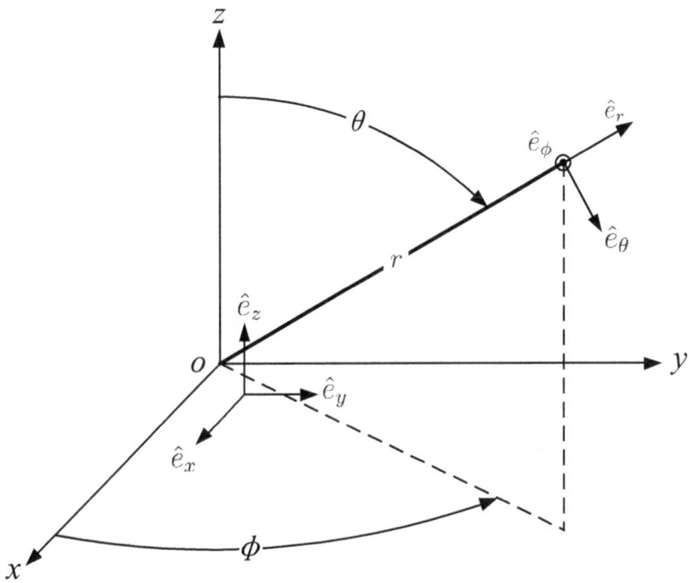

Figure 1.4 Polar coordinate and unit vectors

1.2.2 Loop Antenna

MC are accomplished with the use of loop antennas. A single-turn circular loop antenna is shown in Fig. 1.5 on the x–y plane at $z = 0$. Let a represent the radius of the coil. Let the wire is assumed to be very thin and the current $I = I_0$, where I_0 is a constant [13].

Then the radiated fields of such a loop antenna at an arbitrary point N_0 are approximately expressed under the spherical coordinates with the magnetic field components [13]:

$$\begin{aligned} H_r &= j\frac{ka^2 I_0 \cos\theta}{2r^2}\left[1 + \frac{1}{jkr}\right]e^{-jkr} \\ H_\theta &= -\frac{(ka)^2 I_0 \sin\theta}{4r}\left[1 + \frac{1}{jkr} - \frac{1}{(kr)^2}\right]e^{-jkr} \\ H_\phi &= 0, \end{aligned} \tag{1.2}$$

while the electric-field components [13]:

$$\begin{aligned} E_\phi &= \eta\frac{(ka)^2 I_0 \sin\theta}{4r}\left[1 + \frac{1}{jkr}\right]e^{-jkr} \\ E_r &= E_\theta = 0, \end{aligned} \tag{1.3}$$

where $k = 2\pi/\lambda$.

The signal energy of MC is transmitted in a near-field region, i.e., $kr \ll 1$. With this assumption, the expressions of the fields given by (1.2) and (1.3) can be simplified as [34]:

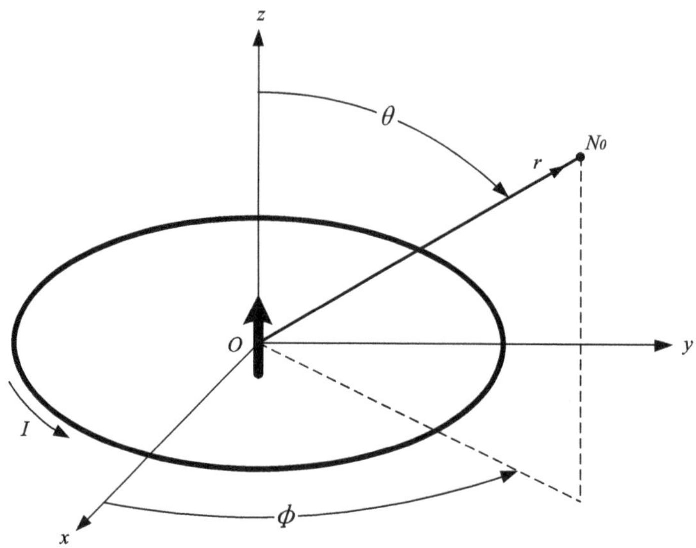

Figure 1.5 Circular loop antenna

$$
\begin{aligned}
H_r &\simeq \frac{a^2 I_0 e^{-jkr}}{2r^3}\cos\theta \\
H_\theta &\simeq \frac{a^2 I_0 e^{-jkr}}{4r^3}\sin\theta \\
H_\phi &= 0
\end{aligned}
\tag{1.4}
$$

$$
\begin{aligned}
E_\phi &\simeq -j\frac{a^2 k I_0 e^{-jkr}}{4r^3}\sin\theta \\
E_\theta &= E_r = 0.
\end{aligned}
\tag{1.5}
$$

1.2.3 Magnetic Moment

Magnetic moment is a fundamental metric to measure the capacity of an MI loop antenna; it is measured by the following equation:

$$m = IA, \tag{1.6}$$

where A represents the area of loop. Additionally, in a standard circular loop, we have $A = \pi a^2$. In this case, the magnetic moment of a loop antenna of N turns is given by

$$m = NIA. \tag{1.7}$$

An MI antenna with a larger magnetic moment can radiate a stronger magnetic field signal.

In an MC system, the magnetic field is measured in the B-field with SI unit tesla (symbol: T). Considering a constant value of transmitting current I, we have the frequency $f = 0$ and $k = 0$, and consequently $e^{-jkr} = 1$. The magnetic field vector at the point N_0 (Fig. 1.5) in a uniform vacuum space is

Table 1.2 Resistivity

Material	$\rho(\Omega\cdot \text{m})$ at 20 °C
Carbon(graphene)	1×10^{-8}
Silver	1.59×10^{-8}
Copper	1.68×10^{-8}
Annealed copper	1.72×10^{-8}
Gold	2.44×10^{-8}
Aluminum	2.82×10^{-8}
Tungsten	5.6×10^{-8}
Iron	9.71×10^{-8}

$$\widehat{B} = \widehat{e}_r \frac{\mu_0 a^2 I_0 N}{2r^3} \cos\theta + \widehat{e}_\theta \frac{\mu_0 a^2 I_0 N}{4r^3} \sin\theta, \tag{1.8}$$

where $\mu_0 = 4\pi \times 10^{-7}$ T · m/A is the permeability of vacuum. Hence, the magnitude of $\widehat{B}$ is

$$\mathbf{B} = \left|\widehat{B}\right| \frac{\mu_0 NIA}{4\pi r^3} \sqrt{3\cos\theta^2 + 1} = \frac{\mu_0 m}{4\pi r^3} \sqrt{3\cos\theta^2 + 1}. \tag{1.9}$$

Equation (1.9) reveals that, for a given range r, the magnetic field strength is proportional to the magnetic moment m, while for a given receiving magnetic field strength threshold, the transmitting range $r \propto m^{1/3}$. We conclude that three ways can be applied to improve the magnetic field signal and the transmitting range from (1.9): enlarging the area A, adding loop turns N, and increasing transmitting current I. Although the three parameters A, N, and I are proportional to the magnetic moment, the consequent increases of consuming power are different.

The power consumed on the antenna loop is directly related with the current I, i.e., $P \propto I^2$. On the other hand, increases in the number of turns N and area A lead to the growth of the direct-current (DC) resistance. A fundamental way to calculate the DC resistance of a loop is Pouillet's law

$$R = \rho \cdot \frac{l_w}{A_w}, \tag{1.10}$$

in which ρ is the resistivity of the loop material, $l_w = 2\pi aN$ represents the total wire length, and A_w is the cross-sectional area of the wire. We provide the resistivity of some conductive materials in Table 1.2. Usually, the antenna loop is made of the wire following the American Wire Gauge (AWG) standard, whose resistance per length ranges from $2 \times 10^{-4}\Omega\cdot \text{m}^{-1}$ to $3\Omega \cdot \text{m}^{-1}$ with different wire diameter). Thus, we can use resistance per length to calculate the total DC resistance:

$$R = \rho_l l_w. \tag{1.11}$$

For a circular loop, $l_w = 2a \cdot \pi$.

After we figure out the calculation of antenna resistance, we can find that the number of turns N is proportional to the DC resistance. On the other hand, the loop area A is related to the circumference as well as the total length: $l_w \propto A^{0.5}$. Given that the

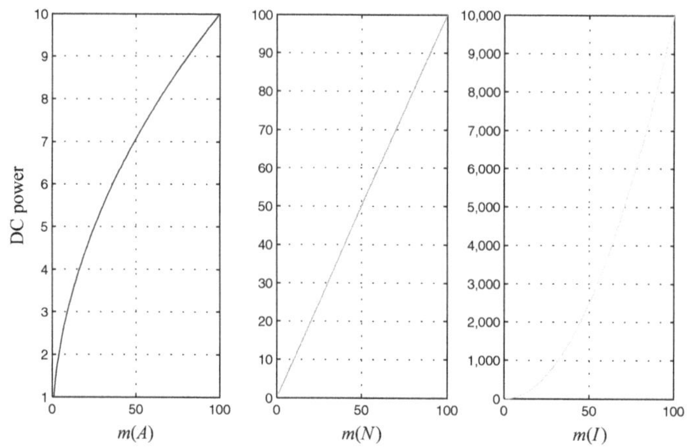

Figure 1.6 Normalized DC power and magnetic moments

power $P \propto R$, we conclude that enlarging the loop area A is the most efficient way to increase the magnetic moment. A normalized DC power versus magnetic moment is shown in Fig. 1.6. The magnetic moments vary from 1 to 100 times by changing A, N, and I, respectively.

It should be noted that only DC resistance and DC power are considered in this section; inductive resistance and conductive resistance in the case of alternating current (AC) will be discussed in Chapter 2. Static magnetic field strength can be treated as a fundamental measurement of the MI signal. The magnetic field sensitivity is a key parameter for a receiver, because advanced MC technologies are all based on the receiving of the magnetic field signal. However, the sensitivity performance of a receiver depends on the technologies of antenna design, antenna manufacturing, and receiving circuit, which is costly and empirical.

1.3 Mutual Inductance Circuit

In order to analyze the communication performance, a mutual inductance model is used as presented in Fig. 1.7. In this model, we take the signal frequency into consideration. This model is able to help us to figure out the power-transmitting process in MC. Similar to the wireless power transport, an MC channel evaluated by an electric voltage is characterized by the following equation:

$$
\begin{aligned}
Z_t &= R_t + j\omega L_t + \frac{1}{j\omega C_t}, \quad Z_t' = \frac{\omega^2 M^2}{R_r + j\omega L_r + \frac{1}{j\omega C_r} + Z_L} \\
Z_r &= R_r + j\omega L_r + \frac{1}{j\omega C_r}, \quad Z_t' = \frac{\omega^2 M^2}{R_t + j\omega L_t + \frac{1}{j\omega C_t}} \\
U_M &= -j\omega M \frac{U_s}{R_t + j\omega L_t + \frac{1}{j\omega C_t}}.
\end{aligned}
\tag{1.12}
$$

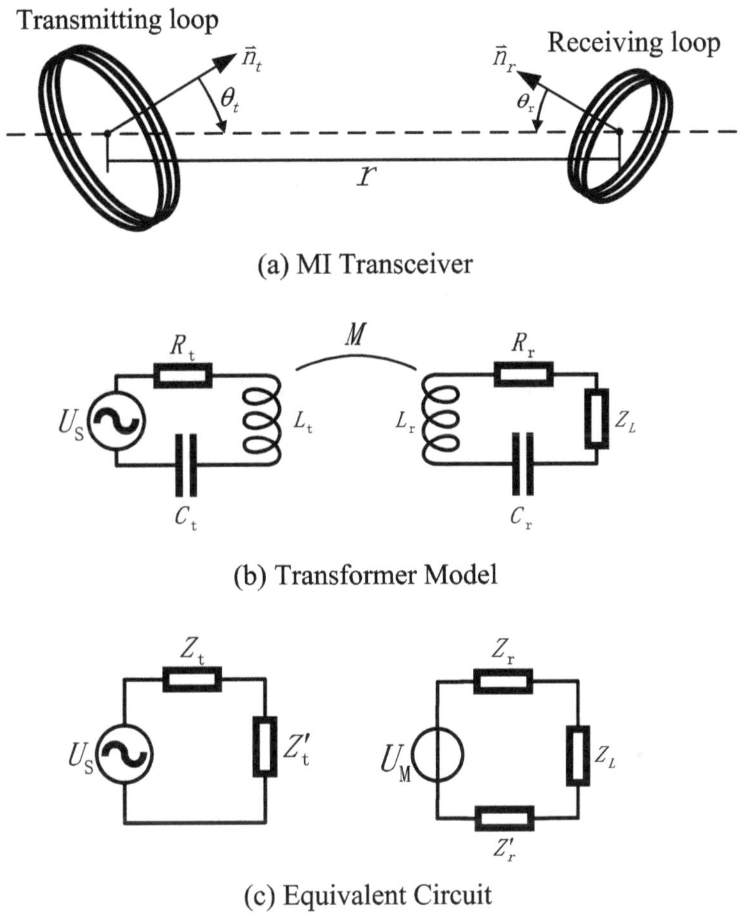

Figure 1.7 Equivalent mutual induction circuit

Here U_M is the induced voltage at the receiving side, which can further be used to derive the receiving power in Section 1.4, U_s is the source voltage at the transmitting side, R_t and R_r are the DC resistances of the transmitting loop and receiving loop, respectively, while L_t and L_r represent the inductance of transmitting and receiving coil, respectively. C_r and C_t are the capacitances that are decided by the resonant signal frequency given as follows:

$$C_r = \frac{1}{(2\pi f)^2 L_r}, \quad C_t = \frac{1}{(2\pi f)^2 L_t}. \tag{1.13}$$

1.3.1 Self-Inductance

According to the definition of inductance, we establish that a current I in the transmitting loop produces a magnetic flux Φ_B through the central region of the loop. With the flux known, the self-inductance is obtained as [14]

$$L = \frac{N\Phi_B}{I}, \tag{1.14}$$

where L is the self-inductance, N is the number of loop turns, Φ_B is the magnetic flux, and I is the current.

The magnetic flux Φ_B is calculated as follows:

$$\Phi_B = nl\mathbf{B}A, \tag{1.15}$$

where l is the length near the middle of the loop, n is the number of turns per length, and A is the area of the loop. The magnitude

$$\mathbf{B} = \mu In, \tag{1.16}$$

where μ is the magnetic permeability of space medium, and from Eq. (1.14), we have

$$L = \frac{N\Phi}{I} = \frac{nL\mathbf{B}A}{I} = \mu n^2 lA. \tag{1.17}$$

For an ideal loop, when $n = N$ and l is set to $0.5a$ [15], the self-inductance of transmitting and receiving loop is

$$\begin{aligned} L_t &\simeq \frac{1}{2}\mu\pi N_t^2 a_t \\ L_r &\simeq \frac{1}{2}\mu\pi N_r^2 a_r. \end{aligned} \tag{1.18}$$

1.3.2 Mutual Inductance

Similar to the calculation of self-induction, we assume that the current I produces a magnetic flux Φ_r through the receiving loop. The mutual inductance of the two loops in Fig. 1.7 is then obtained as

$$M = \frac{N_r\Phi_r}{I}. \tag{1.19}$$

Since the mutual inductance is defined under a static situation, we use equation (1.8) to calculate the magnetic flux Φ_r:

$$\begin{aligned} \Phi_r &= \widehat{B} \cdot \overrightarrow{n}_r \cdot A_r \\ &= \left(\widehat{e}_r \cdot \overrightarrow{n}_r \cdot 2\cos\theta_t + \widehat{e}_{\theta_t} \cdot \overrightarrow{n}_r \cdot \sin\theta_t\right) \frac{\mu a_t^2 IN_t}{4r^3} A_r \\ &= (2\cos\theta_r \cos\theta_t + \sin\theta_r \sin\theta_t) \frac{\mu\pi a_t^2 a_r^2 IN_t}{4r^3}. \end{aligned} \tag{1.20}$$

As a result, we have

$$M = \mu\pi N_t N_r \frac{a_t^2 a_r^2}{4r^3}(2\sin\theta_t \sin\theta_r + \cos\theta_t \cos\theta_r). \tag{1.21}$$

1.3.3 Skin Effect

Because of the changes in the magnetic field, MC are influenced by the skin effect [16]. The alternating magnetic field of the MI system caused an alternating electric current to become distributed with a conductive material, and the electric current flows mainly between the outer surface and a level called the skin depth, δ. The skin effect can be ignored if the operating frequency is low, because the skin depth is very large and the consequent EM field existed anywhere in the medium [16]. However, for the MI system with a high carrier frequency of up to tens of megahertz [15, 17], δ becomes much smaller, and the EM field has enough strength within a short range around the MI loop, which significantly weakens the mutual induction between antenna loops.

In order to model the influence of skin effect, we introduce an addition attenuation factor G to the mutual inductance [16]. The addition attenuation factor G is a function of distance r between the antenna loops and the skin depth δ in the medium. According to the model proved in [18], we have

$$G(r,\delta) = \left| \int_0^{\infty} \frac{x^3}{x + \sqrt{x^2 + j(\frac{\sqrt{2}r}{\delta})^2}} e^{\sqrt{-x^2 - j\frac{\sqrt{2}r}{\delta}}} \cdot dx \right|. \tag{1.22}$$

Let ϵ and σ represent the medium permittivity and conductivity, respectively; then the skin depth δ can by calculated by [19]

$$\delta = \frac{1}{2\pi f \sqrt{\frac{\mu}{\epsilon}\left(\sqrt{1 + \frac{\sigma^2}{(2\pi f)^2 \epsilon^2}} - 1\right)}} \approx \sqrt{\frac{1}{\pi f \mu \sigma}}. \tag{1.23}$$

Although the skin effect is accurately characterized by $G(r,\delta)$, equation (1.22) is not favorable in most applications. Therefore, it has been approximated based on a numerical method by an exponential function $\frac{r}{\delta}$ in [16]:

$$G \simeq 1.004 \cdot e^{-0.1883 \cdot (\frac{r}{\delta})^{1.671}}. \tag{1.24}$$

For the MC channel, the skin effect leads to a decline in the mutual induction between two antenna coils expressed as follows:

$$M = G \cdot \mu \pi N_t N_r \frac{a_t^2 a_r^2}{4r^3} (2 \sin\theta_t \sin\theta_r + \cos\theta_t \cos\theta_r). \tag{1.25}$$

1.3.4 Environment Medium

The MI signal penetrates an underground and underwater lossy medium much more efficiently than EM waves [20]. However, the impact of the medium is non-ignorable. Existing MI research is mainly based on a simple environment, such as a single uniform medium space or an underwater environment with surface reflection and lateral waves [20]. Statistical channels for a complex environment like a Rayleigh fading channel are lacked.

Conductivity

Conductivity measures a material's ability to conduct an electric current. It is commonly represented by the Greek letter σ. The conductivity of a material often varies

Table 1.3 Conductivity

Material	Conductivity $\sigma(S/m)$ at 20 °C
Carbon (graphene)	1.00×10^8
Copper	5.96×10^7
Aluminum	3.50×10^7
Calcium	2.98×10^7
Sea water	4.80
Drinking water	5.00×10^{-4} to 5.00×10^{-2}
Deionized water	5.50×10^{-6}
Silicon	1.56×10^{-3}
Air	3.00×10^{-15} to 8.00×10^{-15}

Table 1.4 Permeability

Material	Permeability $\mu(H/m)$	Relative permeability $\frac{\mu}{\mu_0}$
Vacuum	$4\pi \times 10^{-7}(\mu_0)$	1
Air	$1.25663753 \times 10^{-6}$	1.00000037
Water	1.256627×10^{-6}	0.999992
Concrete (dry)	$4\pi \times 10^{-7}$	1
Aluminum	1.256665×10^{-6}	1.000022
Platinum	1.256970×10^{-6}	1.000265
Wood	1.2566376×10^{-6}	1.00000043
Copper	1.256629×10^{-6}	0.999994

with different factors, including temperature, purity, and concentration of water that contains dissolved salts (the conductivity of some common materials can be found in Table 1.3). In a radio frequency (RF)-challenged environment, the transmission medium is mostly a nonconductive material. However, there can be a nonnegligible level of conductivity due to humidity and mineral substances.

Permeability

Determining the permeability of a coal mine is a complex problem; the relative permeability of coal to gas and water depends on the nature of gas, the operational pressure, and fluid–mineral interactions (the permeability of other common materials can be found in Table 1.4 for reference).

1.3.5 Metamaterial

Metamaterials for EM waves have unusual physical features, such as the negative refraction index, including permittivity ε and permeability μ. To achieve a certain refractive index, metamaterials are carefully built to have a smaller structure feature than the wavelength of the respective EM wave. A negative refraction index is an important characteristic of metamaterials to distinguish them from natural existing materials as illustrated in Fig. 1.8.

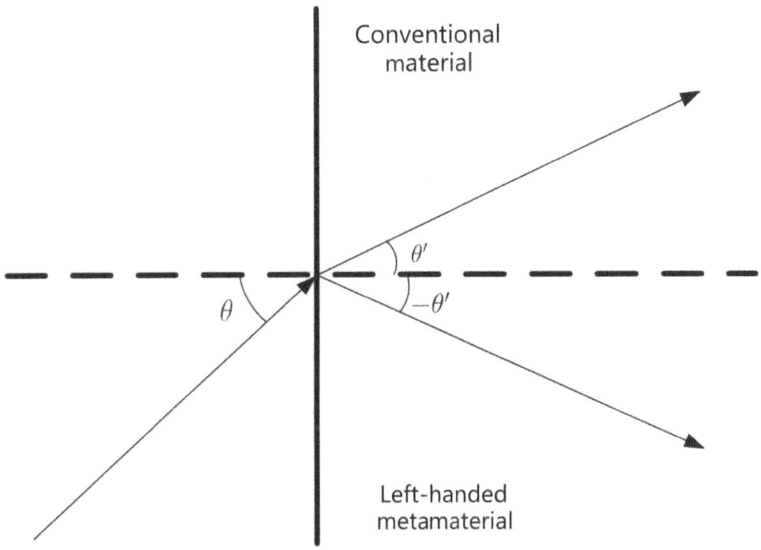

Figure 1.8 Negative refractive index of metamaterial

Application of Metamaterials

Metamaterials are widely used in different kinds of fields. For example, MRI can be enhanced using metamaterials. Long-Term Evolution (LTE) handsets deploy metamaterials for antenna array. Metamaterials can also be deployed in magnetic communications to enhance both the wireless communications using point-to-point MI and MI waveguide. Compared to EM wave-based communication, MI can easily penetrate the lossy medium in RF-challenging environments. One major drawback is the limited transmission distance due to the fast attenuation of magnetic fields. The techniques to enhance magnetic fields using metamaterials will be discussed later.

1.3.6 Waveguide Structure

The transmission distance of an MI system suffers from fast path loss despite its relatively stable channel condition compared to the EM wave. To this end, a waveguide structure using several passive relay devices is employed. The relay point is usually a simple coil that induces a sinusoidal current in the next coil and so on until it passes to the receiver node. Hence, the relay point needs no energy sources or processing devices. The waveguide system is illustrated in [17] and the equivalent circuit diagram of which is shown in Fig. 1.9.

We assume that the waveguide structure uses the same type of coils with the same parameters. To be specific, we let L be the coil self-induction, M be the mutual inductance between adjacent coils, U_t be the voltage of the transmitter energy source, R be the copper resistance of the coil, C be the capacitor loaded in each coil, and Z_L be the load impedance of the receiver. There are totally $(k - 1)$ passive relays that are placed equidistantly between the transceivers.

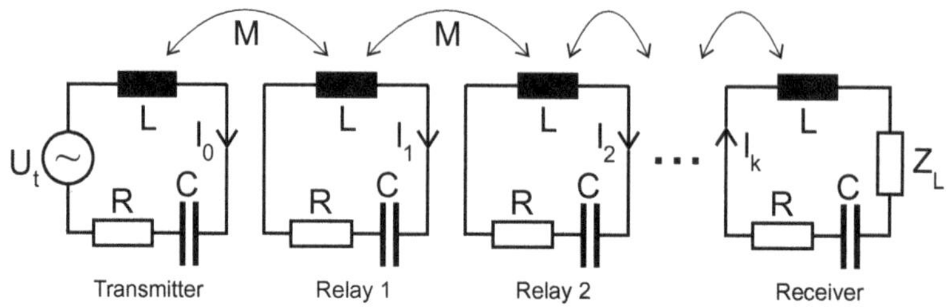

Figure 1.9 Block diagram of an MI waveguide with a transmitter, a receiver, and $(k-1)$ relays

The path loss function is given by

$$L_p(f) = \frac{|S(x,x_L,k) \cdot S(x,x_L,k+1)|}{|Im\{x_l\}|}, \tag{1.26}$$

where f is the signal frequency and j is the imaginary unit, with $x_L = \frac{Z_L}{j2\pi fM}$ and $x_L = \frac{Z}{j2\pi fM}$, and

$$S(x,x_L,n) = F(x,n) + x_L \cdot F(x,n-1), \tag{1.27}$$

with

$$F(x,n) = \frac{\left(\frac{x+\sqrt{x^2-4}}{2}\right)^{n+1} - \left(\frac{x-\sqrt{x^2-4}}{2}\right)^{n+1}}{\sqrt{x^2-4}}. \tag{1.28}$$

The load impedance is matched only to the equivalent impedance Z_e at the carrier frequency using a resistor, i.e.,

$$Z_L = Z_{L,R} = Re\left\{j2\pi f_0 M \cdot \frac{F(x_0,k+1)}{F(x_0,k)}\right\}. \tag{1.29}$$

1.4 MI Communication Performance

1.4.1 Received Power

MC is a near-field technique where communication is achieved by coupling in the non-propagating near-field. The radiation resistance in MC is so small that the radiation power can be neglected. Therefore, the induced power consumed at the MI receiver is the major power consumption. The transmitting power of the MI system consists of the induced power consumed at the MI receiver and the power consumed in the coil resistance. If the antenna impedance is very small, the ratio of the received energy to the propagation energy is 1 because the propagation energy and the received energy are consistent with the distance variation. The advantage of MC is that most of the energy in this technique is delivered to the receiver, and limited energy propagation is wasted in the surrounding environment. Using the circuit model in Fig. 1.7, the transmitted power P_t of the primary coil and the received power P_r are given as follows:

$$
\begin{aligned}
P_t(r) &= \left|\frac{U_s^2}{Z_t + Z_t'}\right| \\
&= \frac{U_s^2}{\left|\frac{C_t M^2 \omega^2}{\frac{1}{jC_t w} + jL_t\omega + R_t} + \frac{1}{jC_t\omega} + jL_t\omega + R_t\right|} \\
&= \left|\frac{M^2\omega^2}{\frac{1}{jC_t\omega} + jL_t\omega + R_t + Z_L} + \frac{1}{jC_t\omega} + jL_t\omega + R_t\right| I_t^2,
\end{aligned}
\tag{1.30}
$$

$$
\begin{aligned}
P_r(r) &= \left|\frac{Z_L \cdot U_M^2}{(Z_L + Z_r + Z_r')^2}\right| \\
&= \frac{M^2 U_s^2 \omega^2 Z_L}{\left|\frac{1}{jC_t\omega} + jL_t\omega + R_t\right|^2 \left|\frac{M^2\omega^2}{\frac{1}{jC_r\omega} + R_r + j\omega L_r + Z_L} + \frac{1}{jC_r\omega} + jL_r\omega + R_r + Z_L\right|^2} \\
&= \frac{1}{2}\frac{|j\omega M|^2 Z_L}{\left|j\omega L_r + \frac{1}{j\omega C_r} + R_r + Z_L\right|^2} I_t^2,
\end{aligned}
\tag{1.31}
$$

where I_t is transmitting current. In the case that MC work at the resonance frequency ($\omega = \omega_0$), $j\omega L_t + 1/jwC_t = 0$, $j\omega L_r + 1/jwC_r = 0$, and $Z_L = R_r$, the transmitted power and the received power are obtained as

$$P_t = \left(\frac{\omega_0^2 M^2}{R_t + R_r} + R_t\right) I_t^2 \tag{1.32}$$

$$P_r = \frac{\omega_0^2 M^2}{4R_r} I_t^2. \tag{1.33}$$

1.4.2 Path Loss

Path loss is the reduction in the power density (attenuation) of an EM wave as it propagates through space or media. It is also a major component in the analysis and design of an MI system. As the distance increases, the receiver will receive less and less energy because of *path loss*. It should be noted that the power is not really lost but not transmitted. The path loss of the MI system with a transmission distance is defined as

$$L_{\mathrm{MI}}(r) = -10 \lg \frac{P_r(r)}{P_t}, \tag{1.34}$$

where "lg" denotes "$\log_{10}$," $P_r(r)$ is the received power at the receiver that is r meters away from the transmitter and satisfy (1.31), and $P_t(r_0)$ is the reference transmitting power when the transmission distance is a very small value and can be looked as (1.30). In what follows, we will discuss the commonly used scenarios.

Under the scenarios of high operating frequency ($R_t \ll \omega\mu$, $R_r \ll \omega\mu$) and considering that r_0 is adequately small, $P_t(r_0) \simeq U_s^2/R_t$. The path loss of the MC system is simplified as

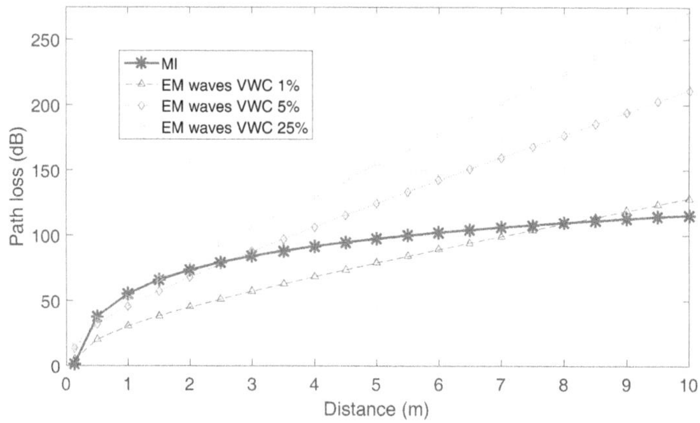

Figure 1.10 Path loss of the EM wave system and that of the MI system with different soil water content

$$
\begin{aligned}
L_{\mathrm{MI}}(r) &= -10\lg\frac{P_r(r)}{P_t(r_0)} = -10\lg\frac{N_r a_t^3 a_r^3}{4N_t r^6} \\
&= 6.02 + 60\lg r + 10\lg\frac{N_t}{N_r a_t^3 a_r^3}.
\end{aligned} \tag{1.35}
$$

The path loss of the MI system and that of the EM wave system are evaluated using MATLAB. For the MI system, the operating frequency f is set to 10 MHz, i.e., $\omega = 2\pi f = 2\pi \times 10^7$; the transmitter and the receiver coils have the same number of turns 5 ($N_t = N_r = 5$) and radius ($a_t = a_r = 0.15$ m). The coil is made of the copper wire with 1.45-mm diameter and with the resistance of unit length $R_0 = 0.01\ \Omega$/m. For the EM wave system, the operating frequency is set to 300 MHz. The permeability of the underground transmission medium is the same as that in the air, which is $4\pi \times 10^{-7}$H/m. In Fig. 1.10, the path loss of the MI system and that of the EM wave system are shown in dB versus the transmission distance with different soil volumetric water content (VWC). It is interesting to see that, compared with the path loss of the EM wave system, the path loss of the MI system is less affected by the earth layer since the permeability is almost unchanged. Therefore, the MI system can achieve smaller path loss than the EM wave system after a sufficient long transmission distance even in the very dry soil medium, which makes MC a promising wireless technology for underground environments.

Let's now consider the scenarios where the operating frequency is low and all antennas of transceivers are identical. Suppose that the resistance of the transmitter and that of the receiver have the same value R, and the MI system works at a resonance frequency ω that is approximately equal to the resonance angular frequency ω_0. According to (1.32) and (1.33), the path loss L_{MI} is then obtained as

$$
L_{\mathrm{MI}}(r) = 10\lg\frac{2(2R^2 + \omega_0^2 M^2)}{\omega_0^2 M^2} \approx 10\lg\frac{4R^2}{\omega_0^2 M^2}. \tag{1.36}
$$

1.4.3 Bit Error Rate and Communication Range

BER is the number of bit errors per unit time. It may be affected by transmission channel noise, interference, distortion, attenuation, wireless multipath fading, etc. Owning to the quasi-static channel of MC, the BER characteristic depends mainly on three factors: the path loss, the noise level, and the modulation scheme used by the system. When the signal level remains the same, the noise level is inversely proportional to the signal-to-noise ratio (SNR), which can be calculated by SNR $= P_t - L_{\mathrm{MI}} - P_n$, where P_t is the transmitting power and P_n is the average noise level. The modulation scheme is an important factor influencing the BER. For example, in the case of QPSK and 2PSK modulation in the AWGN channel, the BER as a function of the SNR is given by BER $= \frac{1}{2}\mathrm{erfc}(\sqrt{\mathrm{SNR}})$, where erfc$(\cdot)$ is the error function.

When BER of the MI link between the transmitter and the receiver increases above a threshold (BER$_{\mathrm{th}}$ = 1%), the communication can be looked as invalid if there are lots of error data that cannot be corrected. Now that the path loss L_{MI} is the function of distance (r) as depicted in (1.35), the fact that the SNR is in relationship with the distance results in BER = BER(r). Therefore, the communication range $r_{\max}$ satisfy

$$\frac{1}{2}\mathrm{erfc}(\sqrt{P_t - L_{\mathrm{MI}}(r_{\max}) - P_n}) = 0.01. \tag{1.37}$$

Equation (1.37) exhibits that the communication range increases with the increase of $\frac{P_t}{P_n}$. As described in Fig. 1.11, where we set P_t as 10 dBm, the transmission range of the MI system is always larger than the EM wave system in a low noise scenario. However, in the high noise scenario, the transmission range of the MI system is between the range of the EM wave system in dry soil and the system in wet soil.

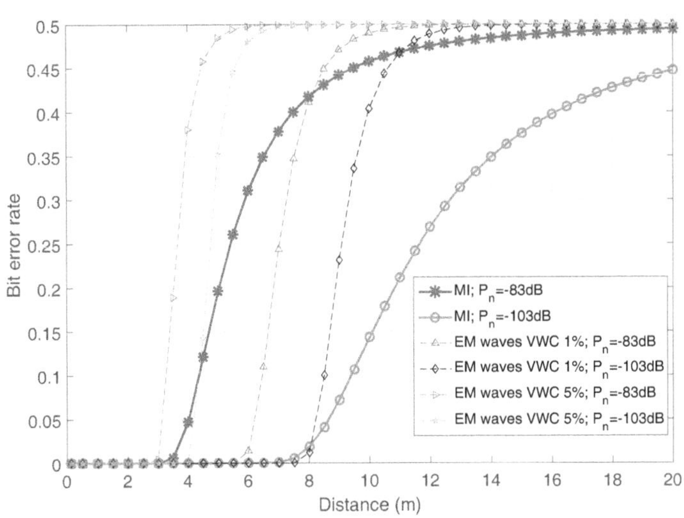

Figure 1.11 Bit error rate of the MI system and the EM wave system with different soil water content and noise level

In practical applications, the operating frequency must be low enough to ensure communicating in the near-field range. For example, we suppose that the operating frequency $f = 10$ kHz, the transmitting power $P_t = 40$ dBm, and the coils of the receiver and the transmitter are identical, the radius $r = 4$ m, and the number of each coil $N = N_t = N_r = 12$. According to (1.37) and (1.36), the propagation distance will be above 600 m.

1.5 Channel Capacity

Despite the numerous advantages, the channel capacity of MC is the primary concerns. With reason that the MI transceivers always work at resonance frequency to ensure the low path loss, the bandwidth of the MI-based channel is much smaller than that of the EM wave-based channel. The narrow bandwidth results in the low capacity. According to the Shannon theorem, the capacity is proportional to the bandwidth and the logarithm of SNR at the receiver. The SNR of the MI system has been introduced in Section 1.4, and the bandwidth will be introduced in the following section.

1.5.1 Bandwidth

The fractional bandwidth of an MI system can be estimated from the loaded quality factors of the transmitter and the receiver by

$$B_w = \begin{cases} \frac{f_0}{Q_1}, & Q_1 > Q_2 \\ \frac{f_0}{Q_2}, & Q_1 < Q_2, \end{cases} \tag{1.38}$$

where f_0 is the resonance frequency, and $Q_1 = \frac{\omega L_r}{R_t + R_r}$ and $Q_2 = \frac{\omega L_r}{R_r}$ are the loaded quality factors of transceivers, respectively. The bandwidth of the MI system can be improved by increasing the resistance and decreasing the self-inductance of the coils.

At this resonance frequency, all the coils can achieve the resonance. However, if a deviation from the resonance frequency occurs, each coil will not achieve the resonant status and the load at the receiver will not match that of the system, and the path loss will increase. As a result, the 3-dB bandwidth B_{MI} is adopted as the MI channel bandwidth. The 3-dB bandwidth can be obtained from the following equation:

$$L_{\mathrm{MI}}\left(f_0 + \frac{1}{2}B_{\mathrm{MI}}\right) - L_{\mathrm{MI}}(f_0) = 3\mathrm{dB}. \tag{1.39}$$

When the transmitter and the receiver have the same number of coils, the 3-dB bandwidth of the MI system is

$$B_{\mathrm{MI}} = \frac{R\left(\sqrt{2}-1\right)}{\mu\pi^2 a N^2}. \tag{1.40}$$

For example, when the operating frequency is 10 MHz, the 3-dB bandwidth of the MI system is around 2 KHz. The bandwidth is not affected by the transmission distance. We can notice that the 2-kHz bandwidth of the MI system is much smaller than that of the EM wave system. The small bandwidth results in the low communication data rate.

1.5.2 Channel Capacity

The capacity of a digital MI system is defined in the light of Shannon's information theory and given by

$$C_a = \int_{f_0-\frac{B_{\mathrm{MI}}}{2}}^{f_0+\frac{B_{\mathrm{MI}}}{2}} \log\left(1 + \frac{P_t(f)L_{\mathrm{MI}}(f)}{P_n}\right), \tag{1.41}$$

where f_0 denotes the resonant frequency, and $f_0 = \frac{1}{2\pi\sqrt{LC}}$, L_{MI} is the path loss of the MI system, P_n is the noise power spectral density, and $P_t(f)$ is the power spectral density of transmit.

The composition of the noise factor includes the external noise and the internal noise. The external noise includes background noise, and co-channel and adjacent channel interference, which has the average power $P_{n,e}$. The internal noise $P_{n,Z_{L,R}}(f)$ is often generated by an amplifier circuit. The internal noise can be simplified as the well-known Johnson-noise power spectral densities $\mathrm{E}\{P_{n,Z_{L,R}}(f)\} = 4KTR$, where $K = 1.38 \times 10^{-23}$ J/K is the Boltzmann constant, T is the temperature in Kelvin, R is the copper resistance in (2), and $\mathrm{E}\{\cdot\}$ denotes the expectation value. In general, the external noise is often between −70 and −110 dBm, and accordingly, $P_{n,e} \gg E\{P_{n,Z_{L,R}}(f)\}\cdot B_{\mathrm{MI}}$.

1.6 Network Connectivity

Apart from the analysis of the MI channel that characterizes the point-to-point performance, in this section, we focus on MI-based networks and start from a fundamental network property: connectivity.

Organized in different types of networks, MI networks can provide required facilities in corresponding scenes. For example, the personal emergency device system provided by the MST company is a one-way downward broadcast network, the primary use of which is to provide a mine-wide emergency messaging and alert system; the Senor network consisted of small MI nodes is organized in an ad-hoc network [15], which has the abilities to gather information from diverse physical phenomena and spread control signals.

Before embarking on the detailed analysis of the MI network connectivity, we introduce several fundamental problems in connectivity analysis in this section. The problems provided in the followings help one understand the framework of connectivity and can be used and extended in MI networks.

1.6.1 System Model

In this part, we introduce two basic factors in the formation of a wireless network. A wireless link describes the connection behavior for an arbitrary node pair, while node deployment decides the location distribution of all nodes in the network. Both factors have important effect on the network connectivity.

Wireless Link

A wireless link between two nodes is decided by whether both of them can reach the other with its wireless signal. The transmission of a wireless signal is mainly influenced by a distance-dependent power decay, along with other effects, including shadowing, fading, polarization gain, *etc*., therefore, we model the wireless link between a node pair χ_n as follows:

$$P_{\chi_n} = \Phi(\chi_n). \tag{1.42}$$

Here P_{χ_n} denotes the probability that the two nodes in χ_n have a connection with each other, Φ is a function of the displacement of two nodes which includes the effects of distance, environments, and polarization.

A basic wireless link model is a uniform range model, which assumes that the signal power decay is only related to the distance. Let r_{χ_n} denote the distance between a node pair χ_n; then the uniform range model is described by the following equation:

$$P_{\chi_n} = \begin{cases} 0 & r(\chi_n) > r_0 \\ 1 & r(\chi_n) \le r_0 \end{cases}, \tag{1.43}$$

where $r(\chi_n)$ is the distance between the node pair and r_0 represents the maximal distance for signal receiving.

For MC, both transmitting and receiving signals have the polarization effect based on the angle between the transmitter and the receiver. Thus, we conclude a basic wireless link model of an MI model as follows:

$$P_{\chi_n} = \begin{cases} 0 & r(\chi_n) > r(f_\theta) \\ 1 & r(\chi_n) \le r(f_\theta) \end{cases}. \tag{1.44}$$

Here the polarization function f_θ depends on the antenna types as well as the specific channel models for given scenarios.

Node Deployment

Besides the node-to-node wireless link, the connectivity performance of the wireless network also depends on the Euclidean distribution of nodes. For large-scale wireless networks, the stochastic geometry-based point process provides a way of estimating the graphic characteristics of the network. The node deployment of the wireless network is assumed either to be deterministic, as the two examples shown in Figs. 1.12(a) and (b), or stochastic as shown in Figs. 1.12(c).

It has been proven that the random point process is a generally effective approach to describe the positions of wireless ad hoc networks [174]. Recalling that the MI

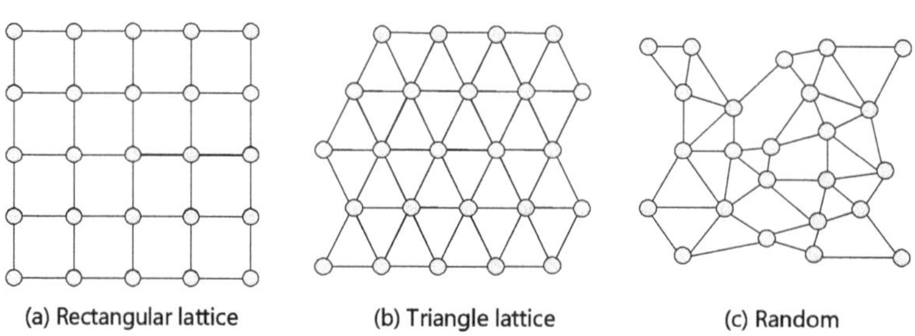

Figure 1.12 Deployment model

networks are proposed for complex environments, which means that the available locations for node deployment are also randomly distributed. As a result, we continue to use the random point process to model the node deployment of the MI network.

The Poisson point process (P.P.P.) is a commonly used deployment model; a homogeneous P.P.P. is defined by the following two properties.

(a) The number of nodes U_d in each finite subspace $\mathbf{D}$ with a size of $||\mathbf{D}|| = D$ follows a Poisson distribution, i.e.,

$$P(n \text{ nodes in } \mathbf{D}) = P(U_d = n) = \frac{\lambda^n}{n} e^{-\lambda}; \; n \in \mathbf{N}_0, \tag{1.45}$$

where $\lambda = \rho D$ represents the expectation $\mathbf{E}\{U_d\}$ and $\mathbf{N}_0$ is the set for positive integers.

(b) The number of nodes N_j in disjoint spaces $\mathbf{D}_j, j \in \mathbf{N}_0$, is an independent random variable, i.e.,

$$P(N_1 = n_1 \bigcap N_2 = n_2 \bigcap \ldots \bigcap N_k = n_k) = \prod_{j=1}^{k} P(N_j = n_j). \tag{1.46}$$

Node Degree

Node degree is the number of neighbors of a node which it can communicate with. Depending on the wireless link mode and node deployment, one can derive the node degree of a certain type of network. The average node degree in a network is an essential factor in the following network connectivity analysis. For a randomly deployed network or nodes with a random wireless link, a node degree is expected.

1.6.2 Connectivity Analysis

Connected Probability

A connected network means that every node in a given network is connected to a single main component (Fig. 1.13) in a way that they can communicate with any other nodes

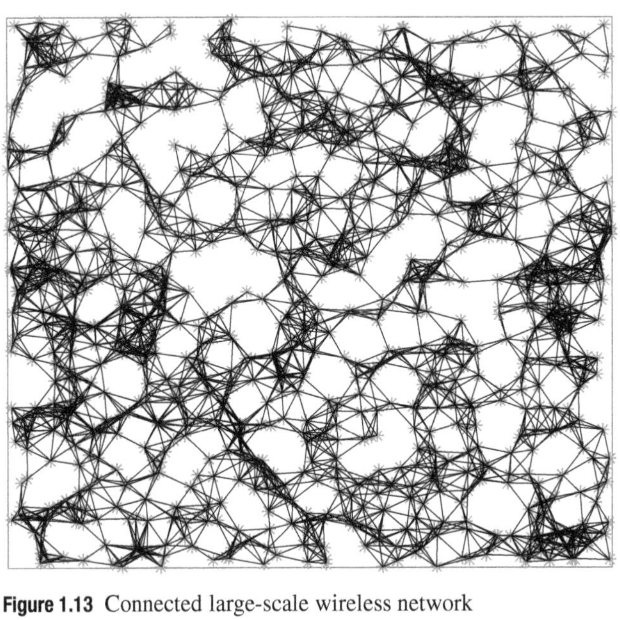

Figure 1.13 Connected large-scale wireless network

in that network. Let **P**(con) denote the probability of a network to be connected. By analyzing the relationship of **P**(con) with the network conditions, including the node number and transmitting power we can find a proper way to establish an effective wireless network in applications. However, if the nodes in a network are randomly deployed as introduced in (1.6.1), it is impossible to find a condition such that the network is surely connected, i.e., $\mathbf{P}(\text{con}) = 1$. Since for random node deployment, the probability for a node to have no neighbor is always positive, we, therefore, use a critical point $\mathbf{P}(\text{con}) = 0.99$ to find the network conditions that make the network *almost-surely* connected.

A more general model for the measurement of a connected network is the *k*-connected network that is defined as follows. There are k independent paths for each node pair in the network to be connected. A connected network is at least 1-connected as shown in Fig. 1.14(b) while an example of 2-connected network is shown in Fig. 1.14(c). Network with a higher *k* has better connectivity, less congestion, and higher robustness.

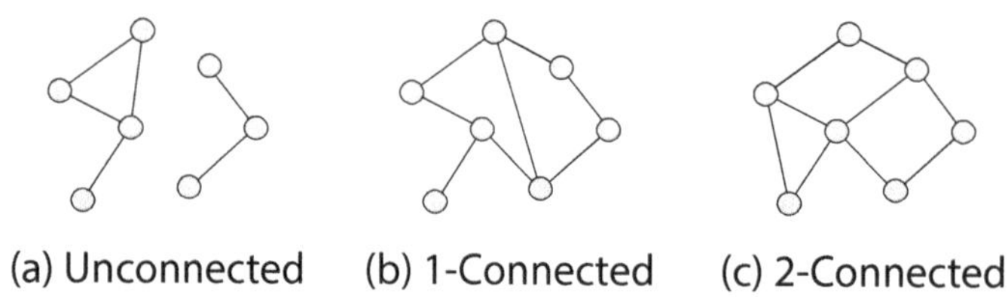

Figure 1.14 *k*-connectivity

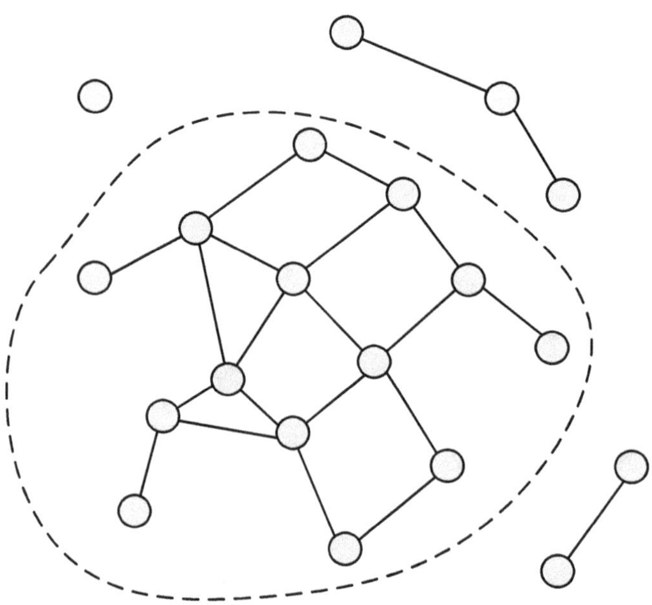

Figure 1.15 Giant component

Giant Component

A giant component of a network is also used to measure a network connectivity. As shown in Fig. 1.15, a giant component is the biggest component in a network that contains the most nodes. By deriving the percentage of the giant component versus the whole network based on the network conditions, one can determine whether the network is connected. A high percentage (close to 1) of a giant component is also a commonly used condition in connectivity analysis.

2 Fundamentals of Magnetic Communications

In this chapter, the differences between MC and EM wave-based communication are summarized and the major advantages of MC are discussed, which provides a big picture of the applicable scenarios of MC. In addition, a physical circuit for MC is introduced. The fundamental performance metrics, such as path loss, bandwidth, capacity, and connectivity, are discussed.

2.1 Magnetic Communication and Other Wireless Communications

With the development of industry, agriculture, and urban infrastructure, more and more underground spaces are exploited. Therefore, there are increasing communication requirements in the underground scenarios. The subway running in the tunnel should communicate with the ground command center. In the automatic agricultural control system, sensors buried in soils need to exchange monitoring data on their temperature and humidity with each other.

Nowadays, several optional communication methods may be used in underground applications with surroundings full of rocks, soils, or water, as discussed in the following subsections.

Electromagnetic Wave-Based Wireless Communication

The EM wave-based wireless communication is widely used in aboveground applications. The main characteristics of the EM wave-based communication can be summarized as follows.

(1) The EM wave-based wireless communication works in ultra high frequency (UHF) or higher frequency bands (above 30 MHz). Therefore, its channel bandwidth and data transmission rate are much higher than other wireless communication methods.

(2) The EM wave-based wireless communication works through the far or radiation field whose magnetic field ($\mathbf{H}$) and electric field E_φ in the spherical coordinate system (d, θ, φ) are

$$\mathbf{H} \simeq \frac{m_d k_0^2 e^{-jk_0 d}}{4\pi d} \sin\theta\hat{\theta}, \tag{2.1a}$$

$$E_\varphi \simeq \sqrt{\frac{\mu_0}{\epsilon}}\mathbf{H}, \tag{2.1b}$$

respectively, where μ_0 denotes the permeability, ϵ denotes the permittivity, m_d is the magnetic moment, and $k_0 = 2\pi/\lambda$ is the wave number, where λ is the wavelength. For the far field range, $k_0 d \ll 1$ is satisfied. The $\mathbf{H}$ and E_φ are in phase and so transmit energy. The propagation medium of the underground communication is no longer air but soil, rock, and water whose permittivity is much larger than air; thus, Eq. (2.1) indicates that EM wave extremely attenuates through these underground media.

Aiming at the underground communication, the EM wave-based communication is often applied in the prosperous area, such as underground malls, subway stations, and urban tunnels. In these area, people can deploy many base stations to extend the communication range. As a result, such a communication method demands high cost in the remote area. In addition, when any disaster strikes, the base stations in underground tunnels may get damaged, and it may become difficult to get them repaired in a short time. Thus, the EM wave-based communication is not a good option for the remote and emergency underground applications.

Long-Wave Communication

Long-wave communication is a wireless communication using EM waves with wavelengths larger than 1,000 m (with frequency lower than 300 kHz), so it is also known as low-frequency communication. Long waves are mainly propagated in the form of ground waves, which can reach a communication distance of tens to hundreds of kilometers on the ground.

Long waves can penetrate rocks and soil, so the long-wave communication is also used for underground communication. The EM waves at the frequency below 60 kHz can penetrate seawater at a certain depth and can be used for underwater communication.

However, the long-wave communication transmitter requires a dramatically large transmit antenna, larger than 10 km. Obviously, such an antenna is difficult to be deployed within the limited underground reachable space. Therefore, the long-wave communication is not a good option for underground scenarios.

SONAR Technology

Sound Navigation and Ranging (SONAR) is a technology that uses the propagation and reflection characteristics of sound waves in water to conduct navigation and ranging through electroacoustic conversion and information processing. It also refers to the use of this technology to detect underwater targets (existence, location, nature, direction of movement, etc.). SONAR equipment for communication is also the most widely used and important device in hydroacoustics. However, sound waves would reflect on the boundary of the medium. Such a phenomenon would bring the remarkable multipath. Therefore, the path loss of SONAR communication in the most underground enurements is quite high. The SONAR technology is not a good option for underground scenarios.

Wired Communication

Wired communication is a method of transmitting information using tangible media such as metal wires and optical fibers. Its channel bandwidth and data transmission rate are higher than most EM wave-based wireless communication. The wired communication is widely used in some underground scenarios, such as in a subway station. However, in the wired communication, the metal wire, pipeline, and other infrastructures need to be deployed in the underground environment. These infrastructures may be significantly deformed by the squeeze of the formation. The construction and maintenance of wired communication for the underground applications are high-cost. Furthermore, these infrastructures may be irreversibly damaged by disasters and are difficult to be repaired in a short time.

Magnetic Induction Communication

The MI communication mainly works by a near-field methods. MI communication is built on the Faradays EM induction law and uses a couple of wired coils to exchange information. Different from the traditional near-field communication, the magnetic fields in soil are additionally attenuated due to the so-called eddy current effect since the propagation medium is electrically conductive. The MI communication mainly works through the near-field methods. In such communication, the transmit antenna(coil) can be treated as a magnetic dipole, and the magnetic and the electric fields at the position of the receive antenna (d, θ, φ) is

$$\mathbf{H} = \mu \frac{n\phi_S^2 I_0}{d^3} e^{-jk_0 d} \left[2\cos\theta(1 + jk_0 d)\hat{e}_d + \sin\theta \left(1 + jk_0 d - (k_0 d)^2\right) \hat{e}_\theta \right],$$

$$\mathbf{E} = \frac{j2\pi f \mu n \phi_S^2 I_0}{4\pi d^2} \sin\theta(1 + jk_0 d)e^{jk_0 d}\varphi \tag{2.2}$$

$$k_0 = 2\frac{\pi}{\lambda} = 2\pi f \sqrt{\mu(\epsilon - j\frac{\sigma}{2\pi f})}.$$

Since the communication distance is within the near field and radiative near-field range, $k_0 d$ is not much larger than 1 (it would satisfy $k_0 d \ll 1$ in most cases). As the transmission distance d increases, the magnetic field strength $\frac{1}{d^3}$ falls off much faster than the EM waves $\frac{1}{d}$ in terrestrial environments. Also, as the permittivity ϵ of the underground medium is much larger than the air and the permeability μ of them is almost the same as the air, the underground medium has little effect on the MI communication but has a significant effect on the EM wave-based communication with far field ($k_0 d \gg 1$). Therefore, compared with the EM wave-based wireless communication, MI is a promising alternative physical layer technique for the WUSN at deep burial depth.

2.2 Mutual Inductance Circuit

2.2.1 Point-to-Point Model

In MI communications, the transmission and reception are accomplished with the use of a coil of wire, which has been modeled in [21]. In other words, the link of the transmitter S and the receiver D can be modeled as a mutual inductance circuit as

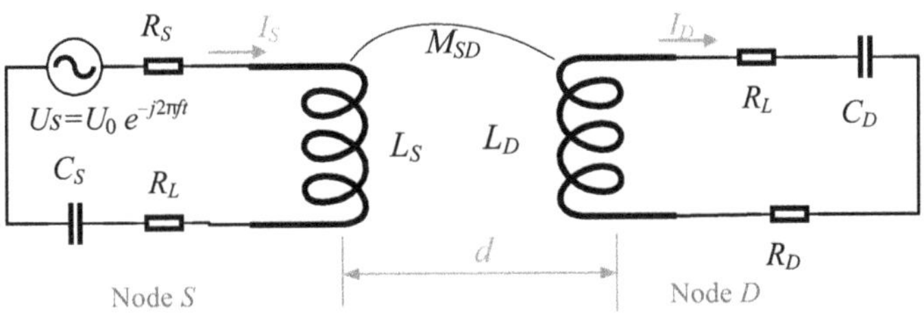

Figure 2.1 Mutual inductance circuit of a point-to-point network

shown in Fig. 2.1, where d is the distance between the coils S and D. In this circuit model, the transmitter and the receiver have the coils L_S and L_D, respectively. They are with the radii of a_s and a_d and with the number of turns n_s and n_d, respectively. Therefore, the inductivities of the transmitting coil and the receiving coil are $L_S = \frac{1}{2}\pi a_s n_d^2$ and $L_D = \frac{1}{2}\pi a_d n_d^2$, respectively. Also, the resistance of the transmitter and that of the receiver coils are $R_S = 2\pi a_s n_s \rho_w$ and $R_D = 2\pi a_d n_d \rho_w$, respectively. Here ρ_w denotes the unit length resistance of the coil of the wire. For each node $i \in \{S, D\}$, a capacity of the capacitor with capacitance C_i need to be tuned to make the circuit resonant at the expected frequency $f_0 = \frac{1}{2\pi\sqrt{L_i C_i}}$, i.e., $C_i = \frac{1}{4\pi^2 f_0^2 L_i}$. Considering R_L the load resistor and f the signal frequency, the total impedances of the transmitter and receiver circuits are

$$Z_S = j2\pi f L_S + \frac{1}{j2\pi f C_S} + R_S + R_L, \tag{2.3a}$$

$$Z_D = j2\pi f L_D + \frac{1}{j2\pi f C_D} + R_D + R_L, \tag{2.3b}$$

respectively.

Suppose the signal powered by the AC voltage U_S in the transmitter coil is a sinusoidal current, i.e., $I_S = I_0 e^{-j2\pi ft}$, where f is the frequency of the transmitting signal. As a result, the receiver coil receives the signal with the current I_D satisfies Kirchhoff's theorem, i.e.,

$$\begin{cases} I_S Z_S + I_D \cdot j2\pi f M_{SD} = U_S, \\ I_D Z_D + I_S \cdot j2\pi f M_{SD} = 0, \end{cases} \tag{2.4}$$

where M_{SD} is the mutual inductivity between the transmitter and the receiver coil. Solve (2.4) and obtain the current I_D:

$$I_D(f) = -\frac{j2\pi f M_{SD} \cdot U_S}{Z_S Z_D}. \tag{2.5}$$

2.2.2 Magnetic Induction Waveguide Model

The MI waveguide model has been widely investigated in many literatures, such as [17], [21], and [22]. The MI waveguide can remarkably improve the performance of the MI communication. A typical MI waveguide structure is applied by deploying $n-1$ relay coils equally spaced along one axis between the transmitter S and the receiver D,

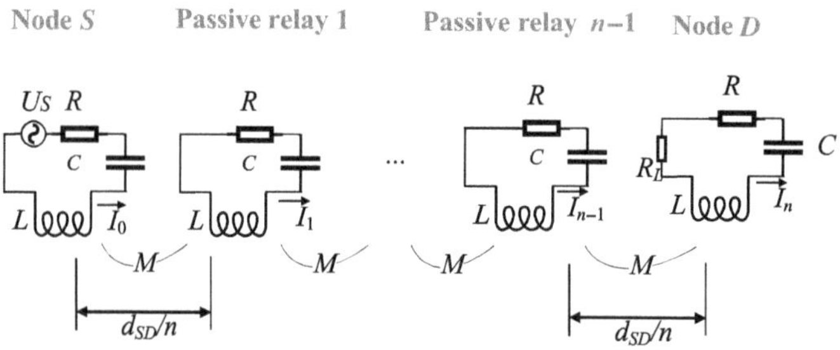

Figure 2.2 Mutual inductance circuit of a point-to-point network.

i.e., the total number of coils is $n+1$. Its mutual inductance circuit is shown in Fig. 2.2, where each node is the same as or similar to the node of the point-to-point model. In this model, suppose the signal powered by the AC voltage U_S in the transmitter coil is a sinusoidal current, i.e., $I_0 = I_0' e^{-j2\pi ft}$, and the current I_k at each coil $k \in [0,n]$ satisfies Kirchhoff's theorem:

$$\left(j2\pi fL + \frac{1}{j2\pi fC} + R\right) \cdot I_k - j2\pi fM \cdot (I_{k-1} + I_{k+1}) = 0,$$
$$\left(j2\pi fL + \frac{1}{j2\pi fC} + R + R_L\right) \cdot I_n - j2\pi fMI_{n-1} = 0. \tag{2.6}$$

Therefore, the current at the receiver coil is

$$I_D = I_n = \frac{U_S}{j2\pi fM} \cdot \frac{1}{\mathrm{Sn}(x, x_L, n+1)}, \tag{2.7}$$

where

$$\mathrm{Sn}(x, x_L, k) = \mathrm{Fn}(x,k) + x_L \cdot \mathrm{Fn}(x, k-1),$$
$$\mathrm{Fn}(x,k) = \frac{\left(\frac{(Z+\sqrt{Z^2-4})}{2}\right)^{k+1} - \left(\frac{(x-\sqrt{x^2-4})}{2}\right)^{k+1}}{\sqrt{x^2-4}}. \tag{2.8}$$

2.2.3 Cooperative Magnetic Induction Communication Model

Cooperative MI communication is to significantly improve the performance of MI communication by an active relay [23, 24, 25]. A typical topology of cooperative MI communication network is shown in Fig. 2.3, where there are a transmitter S, a receiver D and an active relay R. According to the wireless cooperative communication protocol, the signal s generated by a power supplied with voltage U_S is transmitted from S to D in two time slots. During the first time slot, the source broadcasts s to the receiver D and relay R. In this time slot, the mutual circuit can be modeled as shown in Fig. 2.4.

In this circuit model, each node is the same as or similar to the node of the point-to-point model, i.e., each coil has the radius a, the number of turns n, the inductivity

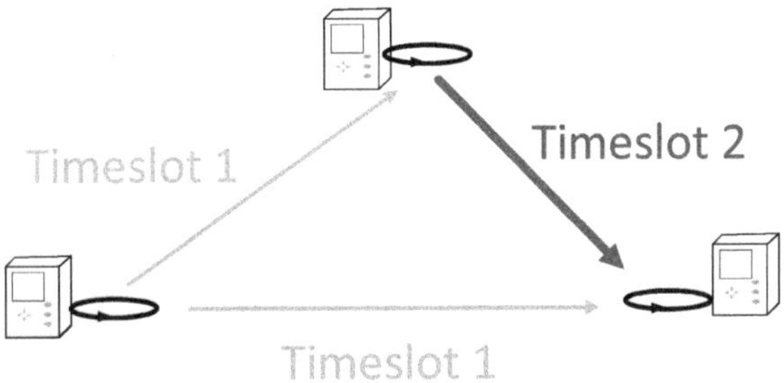

Figure 2.3 A typical topology of cooperative magnetic induction communication network.

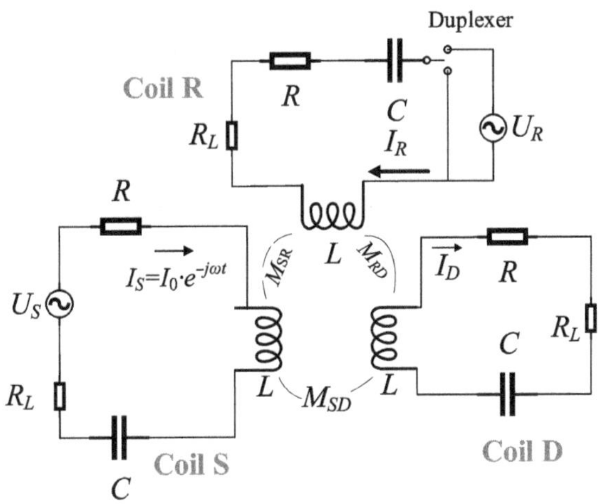

Figure 2.4 Equivalent mutual circuit for the cooperative magnetic induction communication network during the first time slot.

L and the resistance R_0. The capacitor with capacitance C is tuned to make the circuit resonant at frequency $f_0 = 1/2\pi\sqrt{LC}$. R_L is load resistor for minimum power reflections.

Similar to the point-to-point and waveguide models, suppose the signal powered by the AC voltage U_S in the transmitter coil is a sinusoidal current, i.e., $I_S = I_0' e^{-j2\pi ft}$, the current I_k at each coil $k \in \mathcal{N} = S, R, D$ satisfies the Kirchhoff's theorem:

$$
\begin{aligned}
I_S Z_{tr} + \sum_{l \in \mathcal{N} \setminus \{S\}} I_l \cdot j2\pi f M_{Sl} &= U_S, \\
I_S j2\pi f M_{Sk} + \sum_{l \in \mathcal{N} \setminus \{S,k\}} I_l j2\pi f M_{kl} &= -U_{Sk},
\end{aligned}
\tag{2.9}
$$

where $Z_{tr} = Z_{lc} + R_0 + R_L$ in which $Z_{lc} = j2\pi fL + \frac{1}{j2\pi fC}$, $U_{Sk} = I_k Z_{tr}$, $Z_{k_1k_2} = j2\pi f M_{k_1k_2}$ in which $k_1 \in \mathcal{N}$ and $k_2 \in \mathcal{N}$. In most scenarios with the sufficiently large distance and low frequency, the mutual inductances between nodes are relatively small so that $Z_{tr} \gg j2\pi f M_{Sk}$, $Z_{tr} \gg j2\pi f M_{kl}$, $Z_{tr} \gg j2\pi f M_{Sl}$ and $U_S \gg N_k$. Solve (2.9) and obtain the current of each receiver antenna

$$I'_k \simeq \frac{j2\pi f M_{Sk'} Z_{tr}^2 U_S}{Z_{tr}^4}, \tag{2.10}$$

where $k' \in \mathcal{N}\backslash\{S\}$.

During the second time slot, the relay R becomes a transmitter and forward the signal to the receiver D. As the original transmitter S becomes idle, the overall circuit is equivalent to mutual circuit of the point-to-point model.

2.3 Performance Metrics

2.3.1 Path Loss

In the wireless communication field, the path loss (also called propagation loss), refers to the loss caused by the propagation of radio waves in space. It is caused by the radiation diffusion of the transmitted power and the propagation characteristics of the channel, and reflects the change of the average received signal power in the macroscopic range. Therefore, for the MI communication, the path loss can be expressed by the ratio of transmit power to received power, i.e.,

$$Lp = \frac{P_S}{P_D} \tag{2.11}$$

The transmit power P_S and the receive power P_D is written by:

$$\begin{aligned} P_S(f) &= |U_S \cdot I_S|, \\ P_D(f) &= |I_D^2(f)| \cdot Re\{R_L\}, \end{aligned} \tag{2.12}$$

respectively. Here, the current at the load resistor I_D is obtained from (2.5), (2.10), (2.7) and so on.

For the point-to-point model shown in Fig. 2.1, the transmit power and receive power can be re-written by:

$$U_S = |I_S Z_S - Z_{SD} I_D|\,. \tag{2.13}$$

Substituting (2.5), (2.13) and (2.12) into (2.11), the path loss of for the point-to-point link is deduced as

$$Lp = \frac{|Z_D||Z_S Z_D - Z_{SD}^2|}{|Z_{SD}^2| Re\{RL\}}. \tag{2.14}$$

when the communication distance are sufficiently large, $|Z_{SD}| \ll Z_S$ and $|Z_{SD}| \ll Z_D$ are satisfied, so the path loss of most point-to-point link can be expressed by

$$\begin{aligned} Lp &= \frac{|Z_S||Z_D^2|}{|Z_{SD}^2| Re R_L} \\ &= \frac{|Z_S||Z_D^2|}{2\pi f Re\{R_L\}} \cdot \frac{1}{M_{SD}^2}. \end{aligned} \tag{2.15}$$

Due to the mutual inductivity $M_{SD} = \mu \mathbf{H} \cdot \mathbf{n}_D$, where $\mathbf{n}_D$ is the normal vector of the receive coil plane, the path loss of the long-distance point-to-point link is inversely proportional to the sixth power of the communication distance, i.e.,

$$Lp \propto \frac{1}{d_{SD}^6}. \tag{2.16}$$

Similarly, the path loss of the MI waveguide model is deduced as

$$\begin{aligned} Lp_{\text{waveguide}} &= \frac{P_S}{P_D} \\ &= \frac{\text{Sn}(x, x_L, k)\text{Sn}(x, x_L, k+1)}{x_L}. \end{aligned} \tag{2.17}$$

2.3.2 Received Signal-to-Noise Ratio

For any existing communication system, the received SNR is an indispensable performance parameter. As the channel of MI communication using the near field is a quasi-static, the channel power gain that is defined as the inverse of path loss can be used to describe the channel characteristic of the MI link. In other words, the received SNR at the receiver is expressed by

$$\text{SNR} = \frac{P_S}{P_{\text{noise}} \cdot Lp}, \tag{2.18}$$

where P_{noise} is the noise power.

In [26] and [27], Gibson and Kisseleff analyzed the sources of noise in the MC system. For the MI communication network, two sources of noise need to be considered, i.e., thermal noise and ambient noise. As a result, the noise power is given by

$$P_{\text{noise}} = P_{\text{thermal}} + P_{\text{ambient}}. \tag{2.19}$$

Thermal noise, also known as resistance noise, is the noise caused by passive components in MI communication devices, such as coils, resistors, and internal electronic Brownian motion of feeders. It has the characteristics of the Gaussian white noise. According to the well-known Johnson-noise model [28], the power spectral density of thermal noise is given by

$$\mathbb{E}\{U_N^2(f)\} = 4KTR', \tag{2.20}$$

where R' is the copper resistance, $K \sim 1.38 \times 10^{-23}$J/K is the Boltzmann constant, T is the temperature, and $\mathbb{E}\{\cdot\}$ denotes the expectation value. Therefore, according to Kirchhoff's theorem, the current at the receiving coil D is deduced as

$$I_N = \frac{U_N - I_{N,S} Z_{SD}}{Z_D}, \tag{2.21}$$

where $I_{N,S}$ is the induced current by the current I_N. Under most cases, as the communication distance is sufficiently large, $I_{N,S} Z_{SD}$ is much smaller than U_N. Therefore, the average noise power at the receiving coil D is deduced as

$$
\begin{aligned}
P_{\text{thermal}} &= \frac{1}{2} I_N^2 R_L \\
&\simeq \frac{1}{2} \frac{4KTB_w R_D R_L}{Z_D} \\
&= \frac{1}{2} \frac{4KTBR_D R_L}{j2\pi f L_D + \frac{1}{j2\pi f C_D} + R_L + R_D},
\end{aligned}
\tag{2.22}
$$

where B_w is the channel bandwidth.

The ambient noise is determined by the environment of the MI communication. For example, Tan et al. found that the ambient noise in a laboratory environment is about −90 dBm/2kHz [29]. Using a Berkeley Varitronics Systems YellowJacket wireless spectrum analyzer, Li et al. measured an ambient noise of approximately −103 dBm within the sand at 12-inch deep [175]. In the high noise scenario, the ambient noise can reach −83 dBm [30].

2.3.3 Bandwidth

For the analog channel, the bandwidth is the total number of frequencies of the signal that can pass the channel. In the engineering applications, people often use the half-power bandwidth (also called the 3-db bandwidth). The half-power bandwidth is defined as the difference of two frequencies where the received power of the signal drops to the half of the peak value (3 dB lower than the peak value).

Suppose there are a point-to-point MI communications with a sufficiently long communication distance where each node has the same antenna, then the received power is

$$
\begin{aligned}
P_D(f) &= |I_D^2(f)| \cdot Re\{R_L\} \\
&= P_S \cdot \frac{2\pi f Re\{R_L\}}{|Z_D^3(f)|} \cdot M_{SD}^2.
\end{aligned}
\tag{2.23}
$$

For the resonance circuit of the MI antenna, the half-power received power is given by

$$
P_D(f) = P_D(f_0), \tag{2.24}
$$

where $f_0 = 2\pi f \sqrt{L_D C_D}$ is the resonance frequency (center frequency) of the MI antenna circuit. Similar to the method in [24], this equation can be rewritten as

$$
\begin{aligned}
&|Z_D^3(f)| = \frac{1}{2}|Z_D^3(f_0)| \\
&\Rightarrow \left| 2\pi j f L_D + \frac{1}{j2\pi f C_D} + R_D + R_L \right| = 2\left| 2\pi j f_0 L_D + \frac{1}{j2\pi f_0 C_D} + R_D + R_L \right|^3 \\
&\Rightarrow \left| \frac{2\pi j f f_0^2 - 1}{j2\pi f C_D} + R_D + R_L \right|^3 = 2\left| 2\pi j f_0 L_D + \frac{1}{j2\pi f_0 C_D} + R_D + R_L \right|^3.
\end{aligned}
\tag{2.25}
$$

According to the characteristics of the quadratic term, Eq. (2.25) is equivalent to

Table 2.1 BER under the modulation scheme

Modulation	Bit error ratio
BPSK	$\frac{1}{2}\text{erfc}\left(\sqrt{2\cdot\text{SNR}}\right)$
DPSK	$\text{erfc}\left(\sqrt{2\cdot\text{SNR}}\right)$
QPSK	$\frac{1}{2}\text{erfc}\left(\sqrt{2\cdot\text{SNR}}\right)$
$\pi/4$DQPSK	$e^{-2\text{SNR}}\sum_{k=0}^{\infty}\left(\sqrt{2}-1\right)^k \mathbb{I}_k\left(\sqrt{2}\text{SNR}\right)-\frac{1}{2}e^{-2\text{SNR}}\mathbb{I}_0\left(\sqrt{2}\text{SNR}\right)$
MSK	$\text{erfc}\left(\sqrt{2\cdot\text{SNR}}\right)$
M-PSK	$\text{erfc}\left(\sqrt{\text{SNR}}\sin\left(\frac{\pi}{M}\right)\right)$
M-QAM	$2\left(1-\frac{1}{\sqrt{M}}\right)\text{erfc}\left(\sqrt{\frac{3}{M-1}\cdot\text{SNR}}\right)$

Remark: Here, $\mathbb{I}_k$ denotes the k-modified Bessel function of the first kind and $d_{\min}$ is the minimal Euclidean distance.

$$\left|\frac{(2\pi jf)^2 f_0^2-1}{j2\pi fC_D}\right|^3 = 2\left|2\pi jf_0L_D+\frac{1}{j2\pi f_0C_D}\right|^3. \tag{2.26}$$

Solve the equation and obtain two positive solutions f_1 and f_2. The bandwidth is $B_w = |f_1 - f_2|$, i.e.,

$$\begin{aligned} B_w(\mathcal{Z}_C) &\simeq \sqrt{\varpi_w(\mathcal{Z}_C)+\varrho_w(\mathcal{Z}_C)}-\sqrt{\varpi_w(\mathcal{Z}_C)-\varrho_w(\mathcal{Z}_C)}, \\ \varpi_w(\mathcal{Z}_C) &= f_0^2+2\pi^2C_D^2f_0^4\left[\mathcal{Z}_C^{-2/3}-(R_D+R_L)^2\right], \\ \varrho_w(\mathcal{Z}_C) &= \sqrt{\left\{f_0^2+2\pi^2C_D^2f_0^4\left[\mathcal{Z}_C^{-2/3}-(R_D+R_L)^2\right]\right\}^2-f_0^4}, \end{aligned} \tag{2.27}$$

where $\mathcal{Z}_C = \frac{1}{2}(R_D+R_L)^3$. From (2.27), we can find that the antenna position, antenna orientation, and the transmit power have little effect on the MI channel bandwidth.

2.3.4 Bit Error Rate

The BER is a measure of the accuracy of data transmission within a specified time. BER is defined as the ratio of bit errors in transmission to total number of codes transmitted. The BER characteristic depends mainly on the path loss, the noise level, and the modulation scheme used by the system [30]. According to the wireless communication theorem, the average BER of MI communication under different modulation scheme is given in Table 2.1.

From Table 2.1, we find that the BER of MI communications decreases as the SNR increases.

2.3.5 Maximum Communication Distance

The maximum communication distance is the maximum distance between the transmitter and the receiver when the receiver can receive the information with a quality of service (QoS) requirement, including the transmitting rate, delay, jitter, packet error ratio (PER), BER, and outage probability requirements. For instance, for the physical layer, we consider such QoS requirement as the BER requirement. According to Table 2.1, such BER requirement at the receiver node can be mapped into the received SNR threshold $\mathrm{SNR_{th}}$ at the receiver antenna. Thus, the maximum communication distance for the physical layer can be obtained by solving the equation

$$\mathrm{SNR}(d_{\max}) = \mathrm{SNR_{th}}, \tag{2.28}$$

where $d_{\max}$ is the maximum communication distance. For many communication environments, the noise power does not change with time and antenna position, and Eq. (2.28) is equivalent to

$$P_D(d_{\max}) = P_{\mathrm{th}}, \tag{2.29}$$

where P_{th} is the received power requirement. As a result, the expression of $d_{\max}$ with three coils of the tri-directional receiver is [31]

$$d_{\max} = \frac{3\sqrt{2}\mathbf{W}(G_\theta^{1/3}(q_d \cdot q_m)^{1/6} \cdot \omega^{5/6}}{\sqrt{\mu\sigma\omega}}, \tag{2.30}$$

where $\omega = 2\pi f$ is the angle frequency $G_\theta = \sqrt{3\cos\theta_S' + 1}$, where θ_S' is the angle of the observed transmitting direction and the normal vector of the transmitting coil plane, $q_m = \mu^5\sigma^3$ and $q_d = \frac{N_S^2 N_D^2 P_S A_S^2 A_D^2}{6^7\pi^2(P_{\mathrm{th}}R_D)^2}$, where N_S and N_D are the numbers of the turns of the transmitting and receiving coils, respectively, and A_S and A_D are the areas of the transmitting and receiving coils, respectively.

2.3.6 Outage Probability

Outage probability is defined as the probability that a given information rate I is not supported. As the MI channel without mobility nodes is quasi-static, the transmit rate (can be treated as information rate) would not change during the time of the delay requirement. Thus, the outage probability is meaningless. However, the channel of mobility MI communication cannot be treated as a quasi-static since it may contain fast fading caused by the antenna vibration [32], the outage probability becomes an important performance measurement. According to [32], the outage probability of the MI communication is expressed as

$$p_{\mathrm{out}}^{SD}(I) = \mathbb{P}\left[\frac{P_S}{P_D} \cdot |H_{SD}|^2 < 2^I - 1\right], \tag{2.31}$$

where $\mathbb{P}$ denotes the probability, $|H_{SD}|^2 = \frac{1}{Lp}$ is the channel power gain that can be rewritten as

$$
\begin{aligned}
|H_{SD}|^2 &= \frac{1}{Lp} \\
&= \frac{2\pi f Re\{R_L\}}{|Z_S||Z_D^2|} \cdot \frac{1}{M_{SD}^2} \\
&= \frac{2\pi f Re\{R_L\}}{|Z_S||Z_D^2|} \cdot \frac{1}{M_{SD}^2} \\
&= \frac{2\pi f^2 R_L}{|Z_D|^2 |Z_S|} \cdot M_{SD}^2 \cdot J_{SD} \quad \triangleq \aleph_{SD} \cdot J_{SD}.
\end{aligned}
\tag{2.32}
$$

Here M_{SD} is the mutual inductility if all nodes are stationary, J_{SD} is the polarization factor caused by the antenna vibration if there are at least one moving nodes. As J_{SD} is random and called MI fast fading [32], the outage probability is deduced as

$$
p_{\text{out}}^{SD}(I) = \mathbb{P}\left[J_{SD} < \frac{2^I - 1}{\frac{P_S}{P_N}\aleph_{SD}}\right] = F_{J_{SD}}\left(\frac{2^I - 1}{\frac{P_S}{P_D}\aleph_{SD}}\right), \tag{2.33}
$$

where $F_{J_{SD}}$ is the cumulative distribution function (CDF) of the fast fading channel J_{SD}. For the MI communication based on quarter car, the CDF of the fast fading channel is given in [32].

2.4 Channel Capacity

Channel capacity is a widely used metric for the theoretical maximum amount of traffic or signal that can realize a particular infrastructure channel. According to Shannon's theorem, the classic channel capacity formula is given by

$$
C = B_w \log\left(1 + \frac{P_S}{P_N L_p}\right), \tag{2.34}
$$

where B_w is the channel bandwidth, P_N denotes the total noise and interference power (refer to Section 2.3.2), L_p is the path loss, P_S is the transmit power, and $\frac{P_S}{L_p}$ is the received power. For the MI communication, as analyzed in Section 2.3.3, the received power at different frequency is not a constant. Thus (2.34) is rewritten as

$$
\begin{aligned}
C &= \int_{-\infty}^{\infty} \log\left(1 + \frac{P_S}{P_N L_p(f)}\right) df \\
&= \int_{-\infty}^{\infty} C_f(f) df,
\end{aligned}
\tag{2.35}
$$

where $C_f(f) = \log\left(1 + \frac{P_S}{P_N L_p(f)}\right)$ is spectral efficiency. However, the MI system would have an effect on the characteristic of spectral efficiency and the path loss of the overall MI channel. In this section, we introduce the capacities of three typical MI systems as follows.

Point-to-Point System Model

As mentioned in Section 2.3.1, the path loss of the point-to-point link between the transmitter coil S and the receiver coil D can be written as

$$Lp = \frac{|Z_D||Z_S Z_D - Z_{SD}^2|}{|Z_{SD}^2|Re\{R_L\}}.$$

We can substitute this equation into (2.35) and obtain the expression of the capacity of the S–D link:

$$\begin{aligned} C &= \int_{-\infty}^{\infty} \log\left(1 + \frac{P_S}{P_N L_p(f)}\right) df \\ &= \int_{-\infty}^{\infty} \log\left(1 + \frac{P_S}{P_N} \frac{|Z_{SD}^2|Re\{R_L\}}{|Z_D||Z_S Z_D - Z_{SD}^2|}\right) df. \end{aligned} \tag{2.36}$$

For the engineering applications, since the average spectral efficiency outside the bandwidth is much smaller than that within the bandwidth, Eq. (2.38) is approximately expressed by

$$C \simeq \int_{-B_w}^{B_w} \log\left(1 + \frac{P_S}{P_N} \frac{|Z_{SD}^2|Re\{R_L\}}{|Z_D||Z_S Z_D - Z_{SD}^2|}\right) df. \tag{2.37}$$

Also, when the distance between the transmitter coil S and the receiver coil D is sufficiently large, the capacity of the point-to-point MI link is approximately written as

$$C \simeq \int_{-B_w}^{B_w} \log\left(1 + \frac{P_S}{P_N} \frac{|Z_{SD}^2|Re\{R_L\}}{|Z_D||Z_S Z_D^2|}\right) df. \tag{2.38}$$

The Waveguide System Model

Similar to the point-to-point system model, we substitute the expression of the path loss of the MI waveguide model (i.e. (2.17)), we obtain the capacity of the MI waveguide system:

$$C \simeq \int_{-\infty}^{\infty} \log\left(1 + \frac{P_S}{P_N} \cdot \frac{x_L}{\mathrm{Sn}(x, x_L, k)\mathrm{Sn}(x, x_L, k+1)}\right) df. \tag{2.39}$$

3 Magnetic Induction Antennas and Channel Characteristics

One of the important peculiarities of MI communications is the use of magnetic antennas that can process information by using inductive coupling instead of radiation. The magnetic coupling is constrained in the near field, which is more secure since it is not easy to be detected. Moreover, magnetic fields have better penetration efficiency than electric fields. A magnetic antenna is more robust to environment changes than its electric counterpart.

In this chapter, we first introduce the fundamentals of antennas to demonstrate the difference between magnetic antennas and the widely used electric antennas to gradually narrow down our discussion from a big picture. We present the advantages of magnetic antennas and their applicable conditions. Also, we introduce the signal transmission techniques and channel characteristics for the single-input-single-output (SISO) system. Finally, we discuss the multiple-antenna MI system with different antenna placement strategies, which is more reliable and efficient than the SISO system.

3.1 Magnetic Dipole Antennas

Magnetic antennas were originally used in wireless communications because magnetic noises are much weaker than electric noises in free space. Recently, magnetic antennas are extensively used in NFC, wireless power transfer (WPT), and extreme environment communications. In this section, we distinguish the outstanding properties of magnetic antennas for the aforementioned applications and compare them with electric antennas. Moreover, we present the antenna model and its characteristics to provide a rigorous understanding of advantages and limitations.

3.1.1 Magnetic Antenna Properties

Magnetic antennas are widely used in NFC, WPT, and extreme environment communications because of their peculiarities that are not available with electric antennas. There are various types of magnetic antennas, such as dipole, quadrupole, and helical antennas. The magnetic dipole antenna has a low profile and is a predictable and simple model, which is desirable for design and implementation. For the time being, most of the applications using MC adopt a magnetic dipole antenna.

NFC enables secure payment within a short range, which forms a standard character of modern cell phones. NFC requires reasonable received signal strength with several centimeters communication range. To ensure high security, NFC signals have to attenuate dramatically fast as the distance from the transmitter increases to avoid potential attacks. Since the signal strength decreases to a negligible level, attackers cannot detect anything and, thus, the communication channel is secure between the transmitter and the receiver.

When integrated on a cell phone, the radius of an antenna for MC is of several centimeters; robots or wireless sensors may use larger antennas with radii of several tens of centimeters. Compared with their wavelength (more than tens of meters), the antenna is electrically small. As a result, antennas are not efficient in radiation, and therefore most of the power is constrained in the vicinity of the antenna. In other words, if the receiver is very close to an antenna, then detecting signals generated by the antenna becomes challenging. This can satisfy the requirement of NFC. WPT also leverages this property to increase the power transfer efficiency between the transmitter and the receiver since very limited power is radiated into space.

Different from NFC and WPT, the extreme environment communications pursue a long communication range in an inhomogeneous and lossy medium. On the one hand, signals at high-frequency (HF) and lower frequency bands have long wavelength that suffers from a small propagation loss in lossy media. Moreover, compared with electric antennas, magnetic antennas have better penetration efficiency and smaller absorption loss [33]. Both of the two desirable properties make the magnetic dipole antenna a promising solution for wireless communications in extreme environments. On the other hand, the small radiation efficiency is a great enemy of long-distance wireless communications since antennas with small radiation efficiency cannot effectively deliver power to the far field. As of now, there are several solutions to increase the radiation efficiency of magnetic dipole antennas to increase the communication range, such as large radius, multiple turns, and metamaterials, which will be introduced in Section 3.3 and Chapter 4.

3.1.2 Field Model

A magnetic dipole antenna model has been discussed extensively in the literature. Readers are referred to [34] for detailed derivation. Here, we briefly introduce the model and pay more attention to its characteristics that can be related to aforementioned applications. In MI communications, the antenna gain, directivity, and radiation power are not as important as the EM wave-based wireless communications. The range of MI communication is within one wavelength, and there is no radiated EM wave in this region. Therefore, instead of discussing the radiated wave, here we focus on the magnetic field.

The EM fields generated by a single-turn electrically small coil can be solved in spherical coordinates, as shown in Fig. 3.1. The magnetic fields are expressed as a vector $\mathbf{h} = [h_r, h_\theta, h_\phi]$, which can be written as

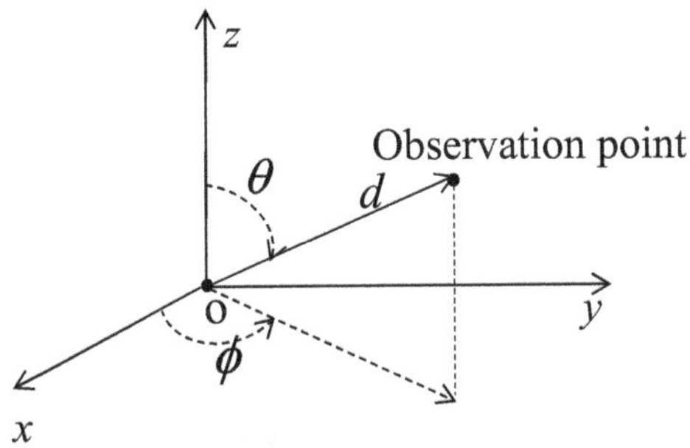

Figure 3.1 Cartesian and cylindrical coordinates system. The origin is a transmit antenna

$$h_r = \frac{jka^2 I_0 \cos\theta}{2d^2}\left[1 + \frac{1}{jkd}\right]e^{-jkd} \tag{3.1}$$

$$h_\theta = -\frac{(ka)^2 I_0 \sin\theta}{4d}\left[1 + \frac{1}{jkd} - \frac{1}{(kd)^2}\right]e^{-jkd} \tag{3.2}$$

$$h_\phi = 0, \tag{3.3}$$

where $j = \sqrt{-1}$, k is the propagation constant, θ is the polar angle, d is the distance from the coil center to the observation point, a is the coil radius, and I_0 is the coil current. Note that the electrically small means the circumference of the coil is much smaller than the wavelength (λ). Usually we consider the circumference is smaller than 0.1λ.

MI communication uses the near-field EM waves, which is different from far-field EM waves. First, the propagation constant is a small number because of the frequency. We can rewrite kd as

$$kd = \omega\sqrt{\epsilon_0\mu_0}d = \frac{2\pi c}{\lambda}\sqrt{\epsilon_0\mu_0}d = \frac{2\pi d}{\lambda}, \tag{3.4}$$

where ϵ_0 and μ_0 are the permittivity and permeability of the air, respectively. In the air, the wavelength of MI signals can be larger than tens of meters, while the communication range is only several meters or centimeters. Thus, kd is much smaller than 1, and (3.1) and (3.2) can be approximated by

$$h_{r,n} = \frac{a^2 I_0 \cos\theta}{2d^3} \tag{3.5}$$

$$h_{\theta,n} = \frac{a^2 I_0 \sin\theta}{4d^3}. \tag{3.6}$$

In contrast, in the EM wave-based wireless communications, we usually adopt the far-field approximation by assuming that the distance is much larger than the wavelength. Under this assumption, (3.1) and (3.2) can be approximated by

$$h_{r,f} = \frac{jka^2 I_0 \cos\theta}{2d^2}e^{-jkd} \tag{3.7}$$

$$h_{\theta,f} = -\frac{(ka)^2 I_0 \sin\theta}{4d}e^{-jkd}. \tag{3.8}$$

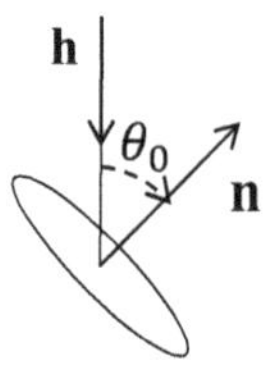

Figure 3.2 Coil orientation and incident magnetic fields

The above equations suggest that using the near-field approximation, the magnetic field decreases as a function of $1/d^3$ for both h_r and h_θ, while using the far-field approximation, the magnetic field decreases as a function of $1/d$ for h_θ (if the distance is large, h_r can be neglected). As a result, the magnetic field generated by the coil antenna decreases dramatically in the near field. If the transmission power is limited, we cannot detect the MI signal in the far field; the communication range is very limited, which can increase the security level.

3.1.3 Receive Mode and Polarization Loss

The coil antenna is linearly polarized that can cause significant polarization loss. First, let us study the receiving mode. The coil antenna is extensively used as a magnetic field probe or a receiving antenna since it receives less electric noises. A single-turn coil cannot receive signals carried by electric fields since the voltage is equivalent to a closed-loop integral of electric fields, which is zero. The induced voltage by the magnetic field across a coil can be written as

$$v = \frac{dB}{dt} = j\omega\pi a^2 \mu_0 \mathbf{h} \cdot \mathbf{n}, \tag{3.9}$$

where B is the magnetic flux density and $\mathbf{n}$ is a unit vector denoting the orientation of the receive coil, as shown in Fig. 3.2. Since $\mathbf{h} \cdot \mathbf{n} = |h||n| \cos(\pi - \theta_0)$, v is affected by the angle θ_0. When $\theta_0 = \pi/2$, the received v is zero, while when $\theta = \pi$ or 0, the received v can be maximized.

The coil orientation can be controlled to optimize the received signal when the position of the wireless device is fixed. However, a slightly change in the coil orientation may strongly affect the received signal strength, which is a significant challenge for MI communications. For practical designs, we have to take this into account. The tri-axis coil can be leveraged to overcome this challenge, which will be introduced at the end of this chapter.

3.1.4 Near-Field Power

Through preceding discussions, we know that MI communications do not leverage the radiated far-field power. Here, on the basis of the EM field analysis, we show that MI communications mainly rely on the reactive power, and it also uses the radiated near field that consists of both reactive power and real radiated power.

An antenna generates real and reactive power. The reactive power is constrained in the near field, while the real power is radiated into the far field. However, there is no clear boundary for the transition from reactive power to real power; there is a radiated near-field region in between [35]. To provide an intuitive understanding, here we use approximations of the complex power intensity. Detailed derivations can be found in [36, 37].

Since the coil generates TE mode, the electric field only has the e_ϕ component, which is

$$e_\phi = \eta \frac{(ka)^2 I_0 \sin\theta}{4d}\left[1 + \frac{1}{jkd}\right] e^{-jkd}, \tag{3.10}$$

where $\eta = \sqrt{\mu_0/\epsilon_0}$. Next, we study the power distribution by analyzing the near field and the far field. First, in the near field and the far field, (3.10) can respectively be approximated by

$$e_{\phi,n} = -j\eta \frac{ka^2 I_0 \sin\theta}{4d^2} \quad \text{and} \tag{3.11}$$

$$e_{\phi,f} = \eta \frac{(ka)^2 I_0 \sin\theta}{4d} e^{-jkd}. \tag{3.12}$$

To calculate the power, we need to perform integrals with respect to the θ and ϕ directions. Under such conditions, only the radial direction of the power exists; other components vanish. As a result, the power in the near field and the far filed can be approximated by

$$P_n = j\eta \frac{\pi k a^4 |I_0|^2}{12 d^3} \quad \text{and} \tag{3.13}$$

$$P_f = \eta \frac{\pi k^4 a^4 |I_0|^2}{12} \tag{3.14}$$

In view of (3.13), the near-field power P_n is reactive, while the far-field power P_f is real. Here, we use approximations by assuming that the distance is either extremely small or large. When the distance falls in the middle, the power is the mix of reactive and real power, which is the radiated near field. Since the range of MI communications is small, the majority of the utilized power is reactive. Since the reactive power cannot propagate, MI communications have a limited communication range.

3.2 Near-Field Signal Transmission and Channel Characteristics

MI communications rely on the near-filed fields instead of EM waves. MI antennas are coupled together, and they can influence each other, while in EM wave-based communications, transmit and receive antennas are independent. In this section, we introduce the wireless MI signal transmission with emphasis on the near field. Also, the channel characteristics in different environments are discussed.

3.2.1 Near-Field Signal Transmission

Near-field signal transmission can also employ typical terrestrial wireless transmission technologies. Here, we present three novel signal transmission schemes that are designed for MI communications, namely, direct antenna modulation [38], three-band MI digital signal transmission communications [39], and magnetic vector modulation [10].

Direct Antenna Modulation

Direct antenna modulation is proposed in [38], according to which this technique is used to increase the MI communication capacity by directly modulating signals in the coil antenna circuit. The coil size is deep subwavelength, which contains a significant amount of reactive power and a large Q factor. The relation between the bandwidth and the Q factor can be written as

$$B \leq \frac{f_0}{Q}, \tag{3.15}$$

where f_0 is the carrier frequency. The channel capacity is proportional to the bandwidth. Therefore, to increase the capacity, we need to increase the bandwidth. However, for MI communications, if we increase the bandwidth, we have to reduce the Q factor, which in turn reduces the coupling between the coils and increases the path loss. As a result, there is a tradeoff between the bandwidth and the coupling strength to maximize the capacity.

The traditional MI modulation uses phase, frequency, or amplitude to modulate information. The typical OOK (on–off keying) modulation transmits a binary logic 1 by sending waves and transmits a binary logic 0 by turning off the source. This method has very limited bandwidth since a large number of cycles are required to fully charge and discharge the coil, which requires long time. The coil itself is an inductor. To make the circuit resonant, capacitors are used to tune the circuit. As a result, the LC circuit is an oscillator when it resonates. The charges and currents oscillate and transform into each other periodically. When the current is at its maximum value, the charge is 0, and all the power are stored in the coil. When the current is 0, all the charges and power are stored in the capacitor. If we use the traditional modulation, the reactive elements in the circuit can store a large amount of power, and thus it takes a long time to charge them. Such a long-time charging and discharging generate significant inter-symbol interference. As a result, the communication data rate is small.

The direct antenna modulation is more efficient in using the bandwidth. It adds a single pole double throw (SPDT) switch in a transmitting coil circuit, as discussed in [38]. To send a binary logic 1, it simply connects to the capacitor and the voltage source to make a resonant RLC circuit. To send a binary logic 0, it connects the coil to the ground to form a circuit that only has a single coil. In this way, the capacitor can always be fully charged. It takes less time to charge the reactive elements by switching at the optimal time.

As discussed in [38], the SPDT switching moment is also important, which can determine the signal quality. There are two optimal switching moments. The first one is when the current in the inductive coil is the maximum, and the second one is when the current in the inductive coil is zero. For example, we switch when the current is zero and all the charges are stored in the capacitor. Now, we want to send a binary logic 0 and connect the coil to the ground. Since the capacitor has no route to discharge, its voltage can be maintained at the proximity of the maximum value. Next, if we want to transmit a binary logic 1, the switch reconnects to the capacitor. Since the capacitor does not lose charges, it only takes a short time to charge it. In this way, the data rate can be as large as the carrier frequency.

Consider that the transmitter circuit has a signal source V_0, source resistance R_s, coil resistance R_1, coil inductance L_1, and a tuning capacitor C_1, which are connected in series, and the receiver circuit has a coil inductance L_2, coil resistance R_2, a tuning capacitor C_2, and a load resistance R_L, which are also connected in series. For the traditional OOK modulation, to transmit a binary logic 1, the current in the transmitter coil is

$$i_{1,t} = \left[i_0 + (i_w - i_0) e^{-\frac{\pi f_0 t}{Q}} \right] \cos(2\pi f_0 t), \tag{3.16}$$

where $i_0 = V_0/(R_s + R_1)$, $Q = 2\pi f_0 L_1/(R_s + R_1)$, i_w is the current at the switching moment, and V_0 is the peak amplitude of the sinusoidal voltage. To transmit a binary logic 0, the current in the transmitter coil is

$$i_{0,t} = i_w \cos(2\pi f_0 t) e^{-\frac{\pi f_0 t}{Q}}. \tag{3.17}$$

For the direct antenna modulation, to transmit a binary logic 1, the current in the transmitter coil is the same as the current when using OOK modulation to transmit a binary logic 1, i.e., $i_{1,\text{dam}} = i_{1,t}$. When transmitting a binary logic 0, the current in the transmitter coil is

$$i_{0,\text{dam}} = i_w e^{-\frac{\pi f_0 t}{Q}}. \tag{3.18}$$

The waveforms for the traditional modulation and the direct antenna modulation are shown in Figs. 3.3 and 3.4, respectively. The red curve is the digital data, i.e., 1 or 0, and the green curve is the waveform. The parameters are given as follows: $i_0 = 1$, $i_w = 0.5$, $Q = 10$ or 20, and $f_0 = 1$ MHz. For the traditional modulation, if the transmit data rate is the same as the carrier frequency, we can barely detect the transmitted symbols correctly since there is almost no difference as shown in Fig. 3.3. On the contrary, by using the direct antenna modulation, even we transmit at the speed of the carrier frequency, we can clearly distinguish the binary logic 1 and 0.

Three-Band MI Digital Transmission

The three-band MI digital transmission is proposed in [39] for underground wireless communications. MI communications have very limited bandwidth as discussed in preceding sections. Therefore, we can transmit very limited data within the bandwidth. To increase the data rate, [39] proposes to use broader bandwidth with adaptive equalization to compensate the significant loss at off-resonant frequencies.

The impulse response of MI communications is very long if we use broadband transmission, which can be up to 10,000 taps. This is mainly due to the high Q of

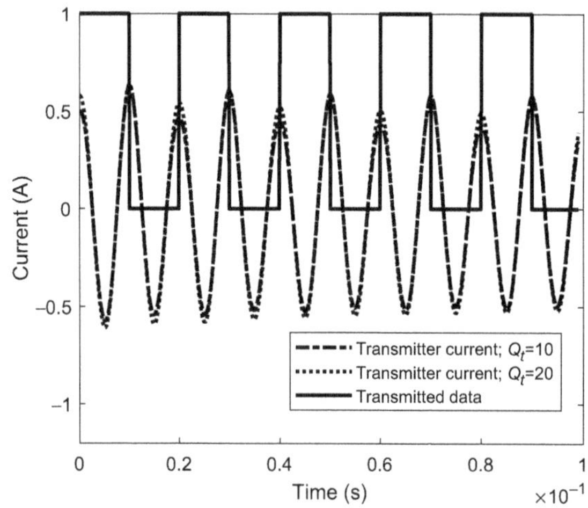

Figure 3.3 Traditional modulation current in the transmitter coil

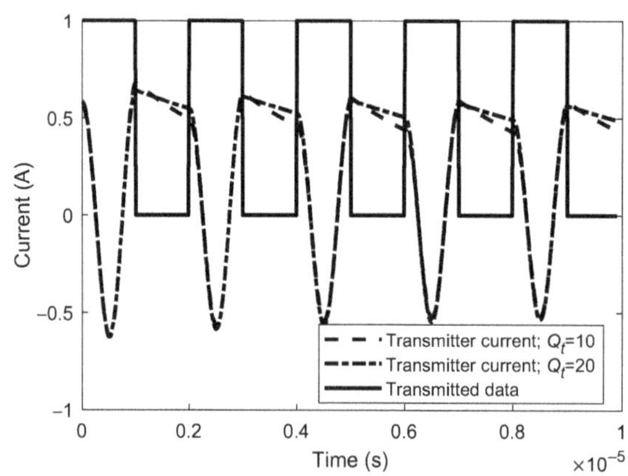

Figure 3.4 Direct antenna modulation current in the transmitter coil

the MI coils that require long time to charge and discharge the reactive elements. Therefore, it is not practical to use a large bandwidth for MI communications.

Assume the SNR with equalization to be

$$\text{SNR}_{\text{eq}} = \frac{|AH_t(f)H(f)|^2}{N(f)(2R_L)}, \tag{3.19}$$

where A is the amplitude of the transmit filter, $H_t(f)$ is the frequency response of the transmit filter, $H(f)$ is the frequency response of the MI channel, and $N(f)$ is the noise power density. The transmit filter can be a root-raised-cosine filter to eliminate the inter-symbol interference. To achieve design objectives, we need a target symbol error rate (SER) associated with the modulation schemes, which is usually set as 0.001. The relation between the SER and the SNR with equalization is

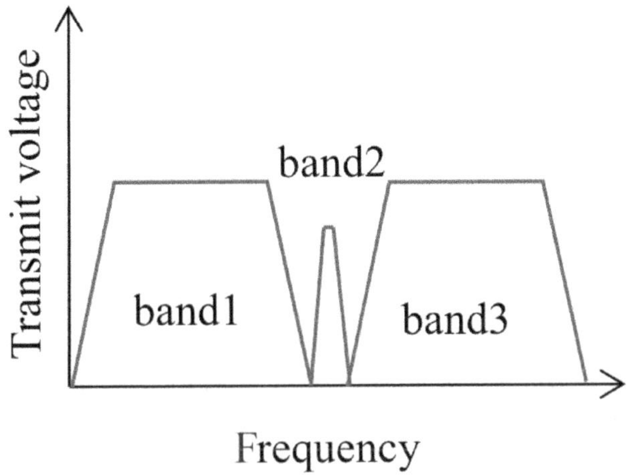

Figure 3.5 Transmit spectra in three-band MI digital transmissions

$$\text{SER} \geq 4Q\left(\sqrt{\frac{3\text{SNR}_{eq}K}{M_{\text{QAM}}-1}}\right), \tag{3.20}$$

where the constellation size of the quadrature amplitude modulation (QAM) modulation is denoted by M_{QAM}, the complementary Gaussian error integral is denoted by $Q(\cdot)$, and the coding gain is K. By using the inverse function of $Q(\cdot)$, we can obtain

$$\log_2 M_{\text{QAM}} = \lfloor \log_2\left(1 + \frac{3\text{SNR}_{eq}K}{c_1}\right)\rfloor, \tag{3.21}$$

where $c_1 = (Q^{-1}(\text{SER}/4))^2$. The operator $\lfloor\cdot\rfloor$ denotes that the number is a power of 2. This could provide a guideline to select the optimum constellation size for reliable communications. The overall data rate is

$$R_d = \frac{R_c \log_2 M_{\text{QAM}}}{T}, \tag{3.22}$$

where R_c is the code rate and T is the symbol period.

Since MI communications have very limited bandwidth, it is a great challenge to transmit the signal with a large bandwidth by using equalization. The channel's impulse response is very long and the computation burden is high. To circumvent this challenge, [39] proposes to divide the wide-band channel into three subbands as shown in Fig. 3.5. By using the subbands, the wide frequency band is divided into several narrow bands, and the equalization requires fewer taps. The reason there are three subbands is that the channel response is symmetrical with the center resonant frequency, i.e., band 2. Band 2 is very narrow and has small path loss due to the resonance. Bands 1 and 3 are wider and possess higher path loss. The frequency band can also be further divided into odd number subbands.

Modulation schemes in different bands can be different. For example, band 2 can support more complicated modulation schemes since its path loss is small, while bands

1 and 3 can only support simple modulation schemes. In order to make bands 1 and 3 support at least 1 bit/symbol modulation, the transmit power has to be optimally allocated.

The data are transmitted in a packet, and each packet has a training sequence for adaptive equalization. The error propagation problem in equalization can significantly increase the symbol errors even the received SNR is reasonable. The problem is especially severe for high-order modulation schemes. To this end, optimizing the SNR is not a good strategy since it cannot guarantee the success of packet transmission. The packet error rate (PER) is used as a design objective to optimally allocate the transmission power.

Let N_p denote the packet length and N_t denote the training sequence length for adaptive equalization; then the total PER can be written as

$$\mathrm{PER} = 1 - (1 - \mathrm{SER})^{N_p - N_t}. \tag{3.23}$$

When there is a packet error, the receiver can only detect the error without correcting it if the transmission is not coded. Hence, the packet has to be transmitted again. This process continues until a packet is correctly received. The effective data rate is no larger than the real data rate. By maximizing the effect data rate, the optimal packet length is given as

$$N_p = \frac{N_t}{2}\left(1 + \sqrt{1 - \frac{4}{N_t \log(1 - \mathrm{SER})}}\right). \tag{3.24}$$

The optimal power allocation strategy is to maximize the effective data rate by using the optimal packet length. Since the three subbands are symmetrical, the power allocated to band 1 is equivalent to that allocated to band 3. The overall transmission power is limited. If we have the knowledge of the power allocated to band 2, the power allocated to bands 1 and 3 can be derived. As a result, the optimal power allocation strategy becomes a search for the optimal power allocated to band 2 to maximize the effective data rate.

Magnetic Vector Modulation

Magnetic vector modulation is proposed in [10] for underground high-speed communications to rescue trapped miners. Although modern underground mines are equipped with wired and wireless communication systems, the earthquake or explosion can destroy these systems and the connections with underground trapped miners. Markham and Trigoni [10] noticed that to penetrate the underground soil and rocks, MC with very low carrier frequency had to be utilized, which resulted in extreme narrow bandwidth with data rate as low as 32 bps. Such low data rate can cause a significant delay in the rescue of the underground trapped miners. Different from direct antenna modulation and three-band digital transmission, magnetic vector modulation uses the orientation diversity provided by the tri-axis coil to improve the communication speed.

The tri-axis coil, as shown in Fig. 3.6, is widely used for the sensing of position and orientation. It is formed by three unidirectional coils with multiple turns and mutually perpendicular orientations. The three unidirectional coils form a complete orthogonal

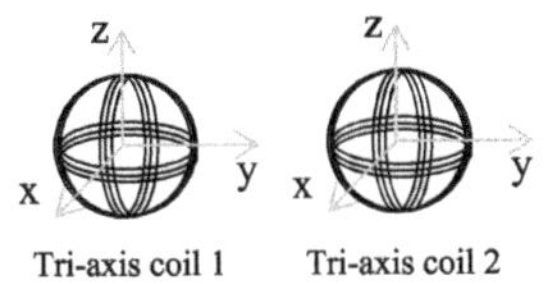

Figure 3.6 Transmitter and receiver with tri-axis coils

coordinate system that can detect the direction of the incident magnetic fields. Also, the transmitter can generate magnetic fields with arbitrary directions in the near field by using excitation currents.

The signal modulation is performed by leveraging the perpendicular coil orientations. In Fig. 3.6, consider that the tri-axis coil 1 is a transmit coil and the tri-axis coil 2 is a receive coil. The modulation scheme is binary phase-shift keying (BPSK). Originally, by using a unidirectional coil, the transmitter can modulate 1 and 0 by sending signals $1 \cdot m(t)$ or $-1 \cdot m(t)$, where $m(t)$ is the baseband waveform. Now, we have three mutually perpendicular coils, and they do not interfere with each other. Thus, the transmitted signal dimension can be extended to three, and the symbols are as follows:

$$\begin{aligned} s_1(t) &= [1,0,0]m(t) \\ s_2(t) &= [0,1,0]m(t) \\ s_3(t) &= [0,0,1]m(t) \\ s_4(t) &= [-1,0,0]m(t) \\ s_5(t) &= [0,-1,0]m(t) \\ s_6(t) &= [0,0,-1]m(t), \end{aligned} \tag{3.25}$$

where the ith element in the bracket denotes the ith coil. As a result, the data rate of using the magnetic vector modulation can be $\log_2 6/\log_2 2 = 2.58$ times larger than the original modulation using the unidirectional coil.

To successfully demodulate the transmitted signals, the receiver requires the knowledge of the channel state information (CSI), which can be derived from channel estimation. Also, from the perspective of the receiver, the six symbols are distinguished, and they can be demodulated uniquely. Consider the two tri-axis coils in Fig. 3.6, if the transmitter x-coil is excited to the transmit symbol $s_1(t)$, the y-coil and z-coil are not activated. If the receiver and the transmitter are well aligned as given in the figure, only the receiver x-coil can detect the signal. Similarly, for the other five symbols, all of them can be uniquely demodulated.

When the coils are not well aligned, the transmitter x-coil sends a symbol, and all the three coils in the receiver can detect the signal. This problem can be easily solved by multiplying the received signal vector by a rotation matrix to virtually adjust the receiver's orientation to make them well aligned.

3.2.2 MI Channel Characteristics

The antenna frequency response and the wireless channel frequency response are independent for the wireless communications using EM waves, since the transmit antenna

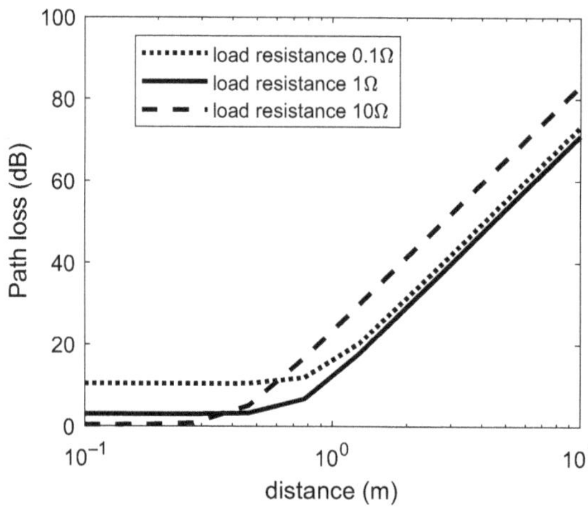

Figure 3.7 Magnetic communication channel path loss in a lossless medium

and the receive antenna are separated by many wavelengths. We cannot directly apply such kind of a model in MC since the transmit antenna and the receive antenna are so close that they are coupled together. The wireless channel and antennas are integrated. On the basis of circuit model, we develop a coil–channel joint model to characterize the frequency response of MC.

Path Loss

A large portion of the link budget is used for the path loss, which can be modeled as the ratio of the received power over the transmitted power. The path loss of the MC channel in a lossless medium is shown in Fig. 3.7. We consider that the coil resistance is 1Ω, number of turns is 5, radius is 15 cm, and the carrier frequency is 10 MHz. The communication medium is considered as air with permeability μ_0 and permittivity ϵ_0. The load and source resistances are the same, which is a variable.

As shown in the figure, the path loss mainly consists of two different regions. In the deep near field, the path loss is almost a constant. Due to the strong coupling, which creates a large reflected impedance, the larger the resistance, the better the impedance matching. Therefore, the largest load resistance, i.e., 10Ω, demonstrates the lowest path loss, and the smallest load resistance, i.e., 0.1Ω, demonstrates the highest path loss. As the distance increases, the mutual coupling between the transmitter and the receiver reduces. As a result, the reflected impedance becomes small, and the dominant impedance is the coil resistance. When the load resistance is matched with the coil resistance, the path loss is the lowest, which is the 1Ω case.

Also, as suggested in the figure, the path loss for MC increases in the speed of 60 dB/decade, i.e., the distance increases 10 times, and the path loss increases 60 dB. This is a good approximation only when the distance is larger than the deep subwavelength region.

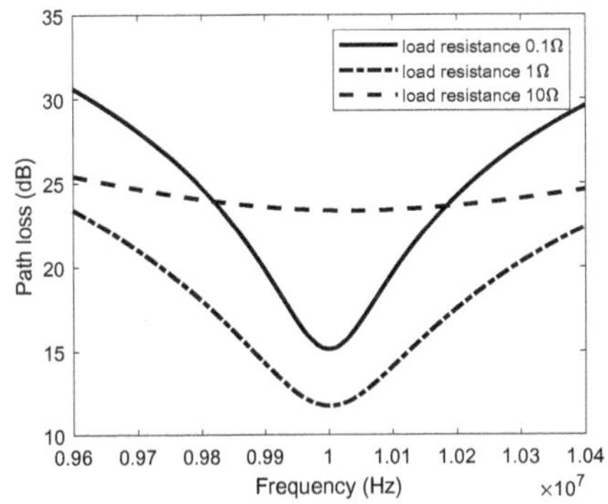

Figure 3.8 Magnetic communication channel frequency response in a lossless medium

Bandwidth

The MC have a narrow bandwidth due to the following two reasons. First, the carrier frequency is small, which fundamentally limits the communication bandwidth. Second, the MC rely on antenna resonance to reduce the large reactive impedance to improve the current in the coil. A strong resonance can significantly improve the signal strength by sacrificing the bandwidth.

By using the same parameter as Fig. 3.7, we plot the frequency response in Fig. 3.8. The high load impedance, i.e., 10Ω, has broad bandwidth since the resonance is weak and the Q factor is small. The smallest load resistance 0.1Ω has the smallest bandwidth but due to the mismatch between the antenna and the load, the path loss is larger than the 1Ω load resistance. The 3-dB bandwidth of MC strongly depends on the coil configurations and is related to the Q factor. There is a trade-off between path loss and the bandwidth and, as a result, there is an optimal configuration to achieve the highest channel capacity.

Magnetic Induction in Lossy Media

Thanks to the high penetrating efficiency, MC are widely used under extreme environments, which are usually conductive, i.e., lossy. In lossless media, it is well known that the self-inductance is a positive real number. However, in lossy media, it becomes complex due to the induced eddy currents in the vicinity of the coil. Since the MC channel is coupled with the antenna and the antenna has very narrow bandwidth, the change of antenna parameters can dramatically affect the communication signal transmission.

Here, we can consider the self-inductance as a complex number, i.e., $L = L_r - jL_i$, where L_r and L_i are the real positive numbers. The imaginary part of L, i.e., L_i can be considered as a frequency-dependent resistance. The eddy current in the vicinity of the coil generates a secondary magnetic field that opposes the primary field, which reduces the current in the MI coil and equivalently increases its impedance. Since the impedance of an inductor is $Z_L = j\omega L$, where ω is the angular frequency, the updated impedance with imaginary inductance is $j\omega(L_r - jL_i) = j\omega L_r + \omega L_i$. Accordingly, the compensation capacitor becomes $C = \frac{1}{\omega_c^2 L_r}$ to achieve the magnetic resonance.

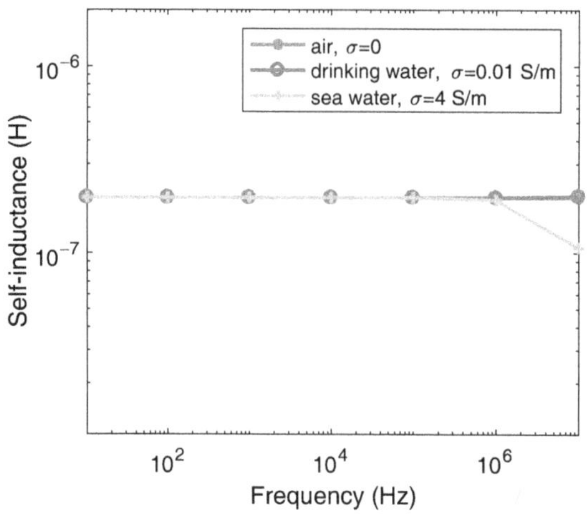

Figure 3.9 Effect of a lossy medium on the real self-inductance

The additional resistance caused by the complex self-inductance can also be treated as additional radiation resistance; the results are consistent with our analysis.

The self-inductance can be calculated as

$$L = \frac{\phi}{I_0} = \frac{\mu \int_S H(0) dS}{I_0}, \tag{3.26}$$

where ϕ is the magnetic flux across the coil surface, I_0 is the coil current, S is the coil area, and $H(0)$ is the magnetic field across the coil surface. According to (3.1), the magnetic field is a function of the propagation constant k. When the medium is lossless, k is a real number, and in the close vicinity of the coil, $H(0)$ is real. As a result, the self-inductance is real. When the medium is lossy, i.e., the conductivity of the medium $\sigma \neq 0$, $k = \omega\sqrt{\mu_0(\epsilon_0 - j\sigma/\omega)}$, which is a complex number. In such a case, the approximations in (3.5) and (3.6) are no longer valid. The reason is that the magnitude of k increases when σ is not zero, which in turn reduces the wavelength, i.e., $2\pi/\Re(k)$. When wavelength is small, kd is not a small number when compared with 1; even d is small than the wavelength in lossless media. In an extreme case, σ is very large, e.g., in sea water σ =4 S/m, $H(0)$ is a complex number when the distance is several centimeters away from the center of the coil.

The effect of the lossy medium on the self-inductance is shown in Figs. 3.9 and 3.10. Three different media are considered, namely, air, drinking water, and sea water, with different conductivity. The single-turn coil is 10 cm in radius. The real part of the self-inductance does not change dramatically with frequency or conductivity. This is the intrinsic parameter of the coil. However, the imaginary part changes significantly, as shown in Fig. 3.10. The equivalent resistance increases with the frequency, and the higher the conductivity, the stronger the effect. For air, there is no obvious effect below 1 MHz. For sea water, the lossy medium creates significant resistance that can consume a large amount of power. This effect has to be well treated to achieve successful communication.

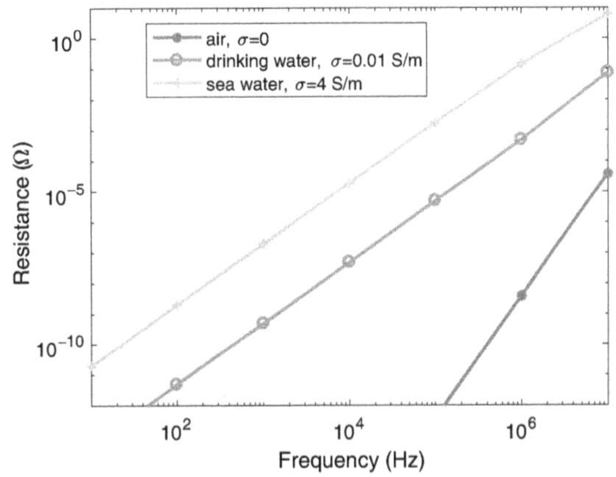

Figure 3.10 Effect of a lossy medium on the overall resistance

3.3 Multi-antenna Magnetic Induction Channel

The magnetic coil is a kind of linearly polarized dipole antenna whose performance is strongly affected by the orientation of the coil. To increase its robustness to orientation change and to provide a reliable wireless communication channel, a multi-antenna solution is used. In a traditional antenna array, the antenna elements are separated by a distance, which is measured in wavelength, and usually the distance is comparable to the wavelength [36]. However, for the MC with a low carrier frequency, one cannot separate the coil with a distance that is comparable with the wavelength, which is longer than tens of meters. If we place the coils in an array close to each other, there is a strong mutual coupling, and the so-called diversity in the traditional antenna array disappears.

Recent works have shown two different methods to increase the diversity. In this book, we classify the method in two categories, namely, the perpendicular multi-antenna [40,41] and the planar multi-antenna [42–44]. The perpendicular multi-antenna array uses coils with different orientations to increase the diversity, which can work for any distance. The planar multi-antenna array uses different coil positions and currents with different phases to increase the diversity, which only works for a short communication range.

3.3.1 Perpendicular Multi-antenna Array

A perpendicular multi-antenna array is a tri-axis coil, which is shown in Fig. 3.6. Three unidirectional coils are mutually perpendicular to each other in orientation to increase the diversity. Originally, a single unidirectional coil can only receive magnetic fields in one direction without polarization loss. Now, three coils can form a virtual orthogonal system and any direction incident magnetic field can be projected onto the three coils. Meanwhile, the three coils can create a magnetic beam by using different currents.

First, we provide a method to generate random orientations of coils with uniform distribution. The orientation of a unidirectional coil is denoted by a unit vector

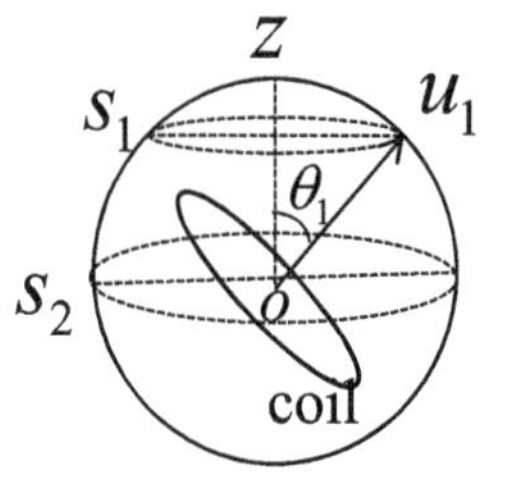

Figure 3.11 Random coil orientation model

$\mathbf{u} = [u^x, u^y, u^z]$, as shown in Fig. 3.11. The end of the unit vector can be randomly placed at any point on a virtual sphere. The way to find the direction of a vector with uniform distribution is as follows.

Consider that the virtual sphere has a radius of 1 m and its origin is at o. The coil orientation is a unit vector $\mathbf{u}$ starting from the sphere's origin o and pointing to u_1 that is a point on the sphere, as shown in Fig. 3.11. Since the orientation is random, u_1 must be uniformly distributed on the sphere. However, the sphere itself is a uniform structure, i.e., the top and bottom are small compared to the equatorial. Therefore, we need to consider such kind of structure.

In order to denote the location using mathematic models, the polar angle θ and the azimuthal angle ϕ are utilized. Since θ and ϕ are independent, the probability of a point u_1 located at (θ, ϕ) is written as

$$P(\theta, \phi) = P(\theta)P(\phi). \tag{3.27}$$

Next, we divide the sphere into slices with infinitely small thickness. Each slice is perpendicular to the z-axis. For example, the slices s_1 and s_2 in Fig. 3.11 are two of these slices. The probability of u_1 being distributed on a slice is proportional to the circumference of that slice which is $2\pi \sin\theta$. Then, $P(\theta)$ and its cumulative distribution function (CDF) can be written as

$$P(\theta) = \frac{1}{2}\sin\theta; \quad \Phi(\theta) = \frac{1}{2} - \frac{1}{2}\cos\theta. \tag{3.28}$$

In order to generate the randomly distributed vectors, the CDF function has to be used. For each slice, the point u_1 is distributed uniformly on the circle with azimuthal angle $\phi = 0$ to $\phi = 2\pi$ and a probability of $P(\phi) = \frac{1}{2\pi}$. Then, we can obtain $P(\theta, \phi) = \frac{1}{4\pi}\sin\theta$. By using the transform relation between spherical coordinates and Cartesian coordinates, the coordinates of $\mathbf{u}$ can be found.

The mathematical model of a tri-axis coil can be represented by three unit vectors since the tri-axis coil is made up of three unidirectional coils. A transmitting coil's orientation and a receiving coil's orientation, in instance, can be written as

$$\mathbf{T} = \begin{bmatrix} t_1^x & t_1^y & t_1^z \\ t_2^x & t_2^y & t_2^z \\ t_3^x & t_3^y & t_3^z \end{bmatrix} \qquad \mathbf{R} = \begin{bmatrix} r_1^x & r_1^y & r_1^z \\ r_2^x & r_2^y & r_2^z \\ r_3^x & r_3^y & r_3^z \end{bmatrix}. \tag{3.29}$$

Here, the subscripts 1, 2, and 3 represent the unidirectional coil numbers.

Moreover, as discussed at the beginning of Section 3.3.1, the three vectors should be mutually perpendicular. Thus, $\mathbf{TT}^T = \mathbf{RR}^T = \mathbf{I}$, where $\mathbf{I}$ is a 3×3 identity matrix. Note that $\mathbf{T}$ and $\mathbf{R}$ are the orthogonal matrices. Without loss of generality, the first vector $\mathbf{t}_1$ can be generated by using the same approach to create a unidirectional coil, i.e., $P(\mathbf{t}_1)$ is the same as a unidirectional coil. Then, $P(\mathbf{t}_2|\mathbf{t}_1) = \frac{1}{2\pi}$, since $\mathbf{t}_2$ should be located on the plane that is perpendicular to $\mathbf{t}_1$. In other words, we consider $\mathbf{t}_1$ as an axis and rotate $\mathbf{t}_2$, the end of $\mathbf{t}_2$ can only be located on a circle with circumference 2π. Once $\mathbf{t}_1$ and $\mathbf{t}_2$ are found, $\mathbf{t}_3$ can be determined. Hence, $P(\mathbf{t}_3|\mathbf{t}_1,\mathbf{t}_2) = 1$. Consequently, the random orientations of the transmit coil antenna and receive coil antenna are captured.

For a randomly orientated tri-axis coil, the radius a and number of turns N are the same for the three unidirectional coils. Since the antenna size is much smaller than the wavelength, the coils can be safely considered as magnetic dipoles. The dipole moment for each coil is $\mathbf{m}_l = m_l \cdot \mathbf{u}_l$, where $l = 1,2,3$, $m_l = \pi a^2 N I_l$, and $\mathbf{u}_l = [u_l^x, u_l^y, u_l^z]$. The dipole moment can be decomposed as $[m_l \cdot u_l^x, m_l \cdot u_l^y, m_l \cdot u_l^z]$. Then, we can add together the three dipole moments along the same direction. For instance, the dipole moment along the x-axis can be written as

$$\mathbf{m}^x = [m_1 u_1^x + m_2 u_2^x + m_3 u_3^x, 0, 0]. \tag{3.30}$$

By using the same procedure, we can obtain $\mathbf{m}^y$ and $\mathbf{m}^z$. We formally replace an arbitrarily oriented tri-axis coil with three unidirectional coils whose axes are the same as the axes of the Cartesian coordinates. We consider the dipole moments $\mathbf{m}^x$, $\mathbf{m}^y$, and $\mathbf{m}^z$ relating to the three equivalent unidirectional coils that are perpendicular to each other. We must rotate the coordinate system and add the magnetic fields as vectors to obtain the magnetic fields produced by coils that are oriented along the x- and y-axes [20].

For a tri-axis coil antenna, the power constraint is $\sum_{q=1}^{3} P_{tq}^o = P_t^o$, where P_{tq}^o is the power allocated to the qth unidirectional coil antenna. Then, we can find that the power density is $P_{tq}(\omega) = P_{tq}^o / B_0$, where B_0 is the signal bandwidth, and the current in an unidirectional coil $I_q(\omega) = \sqrt{2P_{tq}(\omega)/R_c^t(\omega)}$. To make the deductions succinct, we can use the following dipole matrix with decomposition:

$$\mathbf{M}_{mo} = \begin{bmatrix} m_1 & 0 & 0 \\ 0 & m_2 & 0 \\ 0 & 0 & m_3 \end{bmatrix}. \tag{3.31}$$

where $m_x = I_x(\omega)\pi a^2 N, x = 1,2,3$.

As a result, the magnetic field at the observation point can be written as

$$\mathbf{H}_t = \frac{1}{\pi a^2 N} \mathbf{H}_{xyz} (\mathbf{M}_{mo} \cdot \mathbf{T})^{\mathrm{T}}, \tag{3.32}$$

where $\mathbf{H}_{xyz}$ is a 3×3 matrix that consists of the magnetic fields in the x-, y-, and z-directions that are generated by the x-, y-, and z-oriented coils with unit currents. Then, based on the magnetic field, the mutual inductance between the transmit tri-axis coil and the receive tri-axis coil can be written as

$$\mathcal{M} = \mu_0 \pi a^2 N \mathbf{R} \mathbf{H}_{xyz} \mathbf{T}^{\mathrm{T}}. \tag{3.33}$$

Here, $\mathcal{M}$ is a 3×3 matrix that contains the mutual inductance between the three transmit unidirectional coils and the three receive unidirectional coils. Due to the loose coupling, which significantly reduces its value compared to the coil impedance without coupling, we do not take into account the impedance produced by mutual inductance in the analogous circuit. The induced voltage in the receive tri-axis coil can then be calculated based on the mutual inductance,

$$\mathbf{V}_r(\omega) = j\omega\mathcal{M}\mathbf{I}_c, \tag{3.34}$$

where $\mathbf{V}_r(\omega)$ is a 3×3 matrix, and its elements $V_r^{m,n}(\omega)$ denote the voltage in the mth receive coil that is induced by the nth transmitting coil.

Then, the channel capacity can be written as

$$\mathcal{C} \approx \sum_{q=1}^{3} B_0 \log_2 \left[1 + \frac{2\omega_0^2 \Gamma_r(\omega_0) P_{tq}(\omega_0) \lambda_q^2(\omega_0)}{R_c^t(\omega_0)\mathcal{N}(\omega_0)} \right], \tag{3.35}$$

where $\mathcal{N}(\omega)$ is the noise density, $\Gamma_r(\omega)$ includes the antenna impedance matching, $R_c^t(\omega_0)$ is the antenna impedance, $P_{tq}(\omega_0)$ is the transmit power, and $\lambda_q^2(\omega_0)$ is the eigenvalue of $\mathcal{M}\mathcal{M}^\dagger$. In (3.35), we assume that the transmission power density and noise power density are constant within the bandwidth B_0. Based on the developed channel capacity model, the reliability of the tri-axis coil antenna is discussed in the following.

To lessen the effects of the unidirectional coil antenna's erratic orientation, the tri-axis coil antenna is used. From (3.33) to (3.35), we can see the difference between a tri-axis coil and a unidirectional coil with regard to channel capacity. When both the transmit and receive antennas are unidirectional coils, matrices $\mathbf{R}$ and $\mathbf{T}$ would be substituted by the unit vectors $\mathbf{r}$ and $\mathbf{t}$, respectively. Thus, (3.34) can be updated as $v_r(\omega) = j\omega\mu_0\pi a^2 N I_c \mathbf{r}\mathbf{H}_{xyz}\mathbf{t}^{\mathrm{T}}$. When $\mathbf{r} \cdot [\mathbf{H}_{xyz}\mathbf{t}^{\mathrm{T}}] = 0$, we cannot receive any information. In contrast, when $\mathbf{r}/||\mathbf{r}|| = \pm[\mathbf{H}_{xyz}\mathbf{t}^{\mathrm{T}}]/||\mathbf{H}_{xyz}\mathbf{t}^{\mathrm{T}}||$, the receiver can collect the maximum power. Hence, the unidirectional coil antenna is highly unreliable.

Referring back to (3.35), the capacity of the tri-axis coil is directly related to the eigenvalues of $\mathcal{M}\mathcal{M}^\dagger$. Although the rotation of the coil antennas can generate random matrices $\mathbf{R}$ and $\mathbf{T}$, the summation of the eigenvalues of $\mathcal{M}\mathcal{M}^\dagger$ are not influenced by the rotation. By using (3.34), we can obtain

$$\mathcal{M}\mathcal{M}^\dagger = \mu_0^2\pi^2 a^4 N^2 \mathbf{R}\mathbf{H}_{xyz}\mathbf{T}^{\mathrm{T}}\mathbf{T}\mathbf{H}_{xyz}^\dagger\mathbf{R}^{\mathrm{T}}. \tag{3.36}$$

Since $\mathbf{R}$ and $\mathbf{T}$ are the orthogonal matrices, $\mathcal{M}\mathcal{M}^\dagger$ is a similar matrix of

$$\widetilde{\mathbf{W}} = \mu_0^2\pi^2 a^4 N^2 \mathbf{H}_{xyz}\mathbf{H}_{xyz}^\dagger. \tag{3.37}$$

$\mathcal{M}\mathcal{M}^\dagger$ and $\widetilde{\mathbf{W}}$ share the same eigenvalues. More importantly, we find that $\widetilde{\mathbf{W}}$ is independent of antenna orientation. Therefore, by using the tri-axis coil with CSI, we can eliminate the effect of unidirectional coil's random orientation and, ultimately, a stable wireless channel can be formed.

In Fig. 3.12, the distribution of the random coil orientation is shown. The unidirectional coil orientation is denoted by a unit vector that points from the origin toward a

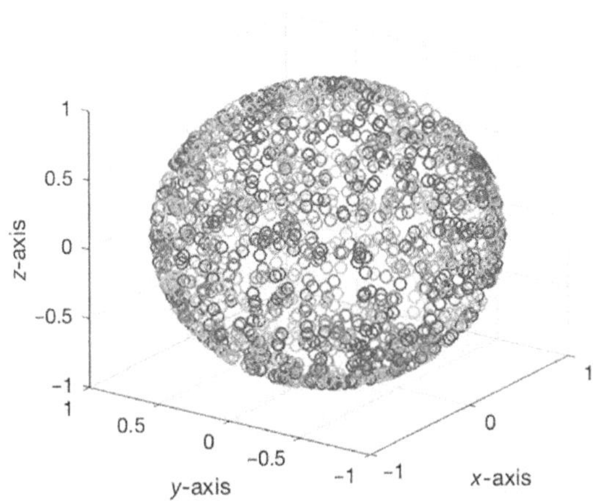

Figure 3.12 Unit vector model to denote the random coil orientation. The vector's end is uniformly distributed on a sphere

point on a sphere. There are 500 random vectors, and their end points are plotted in the figure. As we can see, by using the method in preceding discussions, we can create uniformly distributed vectors.

In Fig. 3.13, we compare the tri-axis coil-based transceivers and unidirectional coil-based transceivers. We randomly generate the coil orientations. The mean channel capacity and the associated standard deviation are plotted in the figure. The wireless channel is assumed to be underwater channel [40]. As we can see, the tri-axis coil-based solution has a very reliable channel without any deviation. However, the unidirectional coil-based solution has a large standard deviation, especially when the SNR is large. In addition, when the tri-axis coil-based transmitter has the CSI, it can optimally allocate power, and its performance is better than that without the CSI at the low SNR regime.

3.3.2 Planar Multi-antenna Array

A planar multi-antenna coil array provides another option to increase the diversity. An example of the planar multi-antenna configuration is shown in Fig. 3.14. By placing multiple coils spatially, the receiver can receive diverse magnetic fields with different phases and magnitudes. For example, the transmitter can provide more power to the transmit coil that has the strongest coupling with the receive coil. Also, if the receive coil orientation is perpendicular to a transmit coil, the remaining transmit coils can provide a nonzero coupling due to the spatial diversity of the coil array. However, this method can only work in a short range. When the distance between the transmit array and the receive coil is large, all the transmit coils have a similar coupling with the receive coil because the spaces between the transmit coils are deep subwavelength. As a result, the diversity disappears when the distance is large. In the following part,

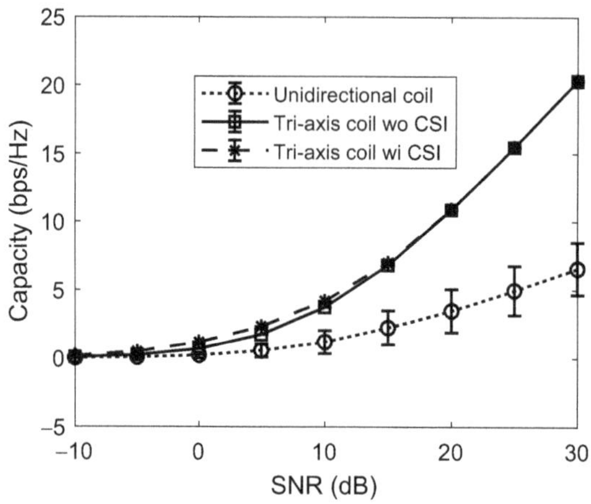

Figure 3.13 Channel capacity of tri-axis coil-based transceivers and unidirectional coil-based transceivers

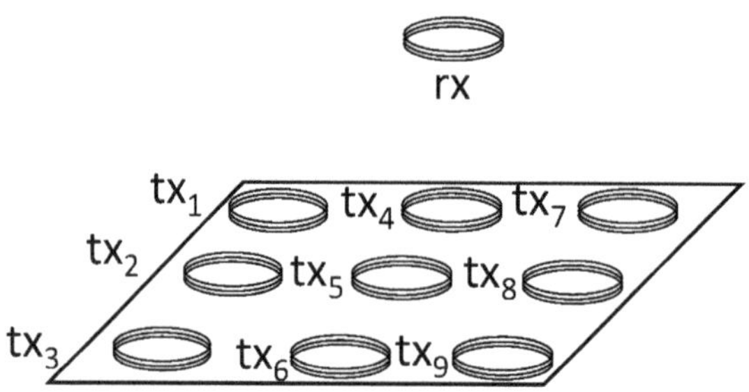

Figure 3.14 Illustration of a planar coil antenna array

we provide the details of a planar multi-antenna channel model with multiple-input multiple-output (MIMO) systems.

Note that some of the content presented here is dedicated for WPT using MI. The relation between MC and WPT is summarized in [45]. One can consider that the MC channel has a wider frequency band than the WPT channel using MI. They share the same fundamental physical principles.

Magnetic MIMO with a planar coil array was first used in WPT to charge portable devices as discussed in [42]. The transmit coils are placed under a desk forming a planar array. The receive coils in portable devices are located around the desk. The range is limited within several meters. The system is robust for random receive coil orientation, and the power consumption is low. By using the magnetic MIMO, the transmitter can generate virtual magnetic beams that can be steered toward the receiver to improve the efficiency.

Assume that there are N_t transmit coils in a planar array. According to the equivalent circuit model, we have

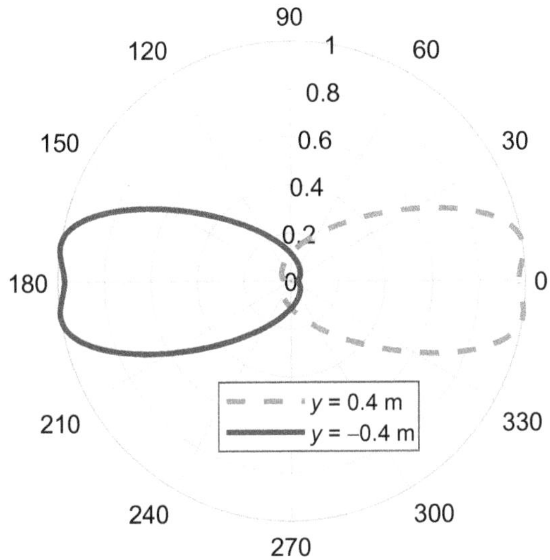

Figure 3.15 Magnetic beams created by a planar array when the distance is small

$$i_l\left(R_l + R_2 + j\omega L_2 + \frac{1}{j\omega C_2}\right) + j\omega\sum_{i=n}^{N_t} M_n i_n = 0. \tag{3.38}$$

By tuning the circuit to resonance, the reactive impedance can be eliminated. Then, we can obtain

$$i_l = \sum_{i=n}^{N_t} \frac{-j\omega M_n}{R_l + R_2} i_n = \sum_{i=n}^{N_t} m_n i_n. \tag{3.39}$$

The term m_n can be considered as a scaled version of the magnetic inductance between the transmit coil and the receive coil. This was referred to as the magnetic channel between the transceivers in [42]. Since the received power is proportional to $|i_l|^2$, we can maximize the received power by maximizing the current in the receive coil. Given the limited transmission power and based on the optimal beamforming strategy, the optimal current in each transmit coil can be written as $i_n = \beta_n i_s$, where

$$\beta_n = \frac{m_n^*}{\sum_{i=1}^{N_t} |m_n|^2}, \tag{3.40}$$

where m_n^* is the complex conjugate of the magnetic channel m_n.

In Fig. 3.15, we compare the performance of the multi-antenna planar array when the distance is small. A planar linear array with five coils is employed, with each coil having five turns and radius 5 cm. The adjacent coils are separated by 20 cm. The center of the coil is set as the origin. A receiver is placed 0.5 m above the array, and the horizontal coordinates of the receiver are 0.4 m and –0.4 m. Then, we plot the generated magnetic field normalized by the maximum value. As can be seen from the figures, when the distance is small or comparable with the array size, the magnetic beams can be a controller to steer toward the receiver.

There are more general analysis of the multi-antenna array with planar structure using spatial diversity in [43, 44]. The optimum transmission algorithms are devised by considering various constraints, such as transmission power, voltage, and current.

Instead of using the spatial diversity, there is another approach to analyzing a multi-antenna planar array using frequency diversity [46]; coils in such a planar array have different resonance frequencies. By leveraging the metamaterial concept [47], the coils can mutually support each other in generating strong magnetic fields.

4 Metamaterial-Enhanced Magnetic Communications

The path loss of MC in the near field increases with the distance of the order of 6 [21], which is much faster than that of using EM waves. In preceding chapters, we have shown reasons by analyzing the power generated by the magnetic dipole antenna and devising the path loss model by using the equivalent circuit model. Due to the high path loss, the MC range is very limited. On the one hand, this can be leveraged to enable secure short-range wireless communications, e.g., NFC. On the other hand, MC cannot be used for many important applications that require a long communication range.

To address this challenge, two solutions are broadly used to increase the MC range. The first solution uses a magnetic waveguide to enhance the generated magnetic fields. A large number of coils are placed with optimal intervals along the propagation path to reduce the path loss. The second solution uses the metamaterial sphere to increase the radiation resistance to improve its radiation efficiency and provide impedance matching with the source or the load.

In this chapter, we first introduce the magnetic waveguide, which starts from the fundamental analysis of its structure and magnetic field propagation in the air. Then, we extend the discussion to extreme environments and show the range extension. Next, we introduce the metamaterial-based solutions. The spherical metamaterial resonance structure is analyzed using advanced EM theory. After that, we present an approach to implementing the spherical metamaterial structure. The enhancement of communication range is demonstrated by using numerical analysis and experimental measurements.

4.1 Magnetic Waveguide

The creation of artificial metamaterials, which have remarkable negative permeability and permittivity, has fundamentally altered the way that materials are designed [48]. Motivated by this, the authors of [49] observed that one could create a waveguide to allow the magnetic wave to travel in the desired direction rather than a 3D material. In this manner, magnetic sensing and magnetic signal transmission can be made use of with guided magnetic waves.

Figure 4.1 provides an illustration of a magnetic waveguide. A transmitter, a receiver, and a number of relays make up its three primary parts. They are all constructed using magnetic coils that are the magnetic dipole antennas, as introduced

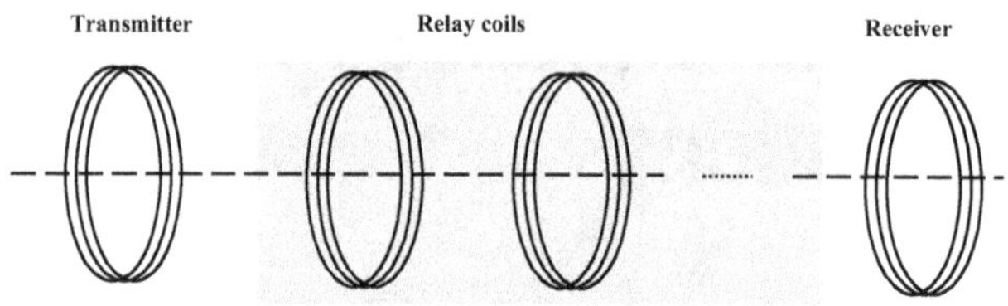

Figure 4.1 An illustration of magnetic waveguide

in earlier chapters. Magnetic fields carry the signals that the transmitter produces. Signals are picked up by the receiver, which then demodulates the data. There is no active input provided by the relays; they are completely passive. The carrier frequency can be changed using the capacitors.

The magnetic waveguide is treated similarly to the transmission line, where a current wave is propagated [49]. For a comprehensive explanation, readers might refer to [50]. Here, we employ the equivalent circuit model and the ideal transformer model [21] in place of taking the propagating wave into account. Next, we suppose that there are $N-2$ relays, one transmitter, and one receiver, which are situated along a straight line. Coils 1, N, and 2 to $N-1$ are used by the transmitter, receiver, and relays, respectively. M_{ij}, where i and j are the coil sequence numbers, is the mutual inductance between two coils.

For the equivalent circuit model, we obtain

$$Z = R + j\omega L + \frac{1}{j\omega C} \tag{4.1}$$

$$Z_{i(i-1)} = \frac{\omega^2 M^2}{Z + Z_{(i+1)i}}, i = 2, 3, \ldots, N-1 \tag{4.2}$$

$$Z_{N(N-1)} = Z_L \tag{4.3}$$

$$Z_{(i-1)i} = \frac{\omega^2 M^2}{Z + Z_{(i-2)(i-1)}}, i = 3, 4, \ldots, N \tag{4.4}$$

$$Z_{12} = \frac{\omega^2 M^2}{Z} \tag{4.5}$$

$$U_{Mi} = -j\omega M \frac{U_{M(i-1)}}{Z + Z_{(i-2)(i-1)}}, \tag{4.6}$$

where $Z_{i(i-1)}$ is the influence of the ith coil on the $(i-1)$th coil, and U_{Mi} is the induced voltage on the ith coil. As a result, the received power at the receiver can be calculated as

$$P_r = \Re\left\{\frac{Z_L U_{Mn}^2}{(Z_{(n-1)n} + Z + Z_L)^2}\right\}. \tag{4.7}$$

The transmitted power is

$$P_t = \Re\left\{\frac{U_s^2}{Z_t + Z_{21}}\right\}. \tag{4.8}$$

Note that, in the above analysis, we implicitly assume that only the neighboring coupling is effective, i.e., the nonadjacent coils have no coupling. In actuality, the non-coupling may take effects, which depends on the distance between the coils. On the basis of the received power and the transmitted power, we can find the path loss, which is $l = -10\log_{10}(P_r/P_t)$.

4.2 Spherical Metamaterial-Enhanced Magnetic Communications: Theoretical Analysis

Although the magnetic waveguide can extend the communication range significantly, it requires very high density of relay coils, which prevents practical implementation. To this end, we introduce metamaterials to MC, which can manipulate and enhance the magnetic fields transmitted and received by magnetic transceivers [51]. In this section, we first provide an overview of the metamaterial development and then introduce the theoretical foundations of metamaterial-enhanced magnetic communications (M^2I). The implementation issues are introduced in the following section.

4.2.1 Metamaterials and Design Challenges

The metamaterial was suggested in [52], where Veselago considered that a specific material with negative dielectric characteristics could provide various propagation pathways for EM waves. Although unusual phenomena have been predicted, it was unclear how to develop and create such materials because they do not exist in nature. Researchers discovered at the close of the twentieth century that carefully constructed artificial structures can produce rare physical phenomena, including backward waves and negative refraction index, which were the same as those predicted by Veselago [53]. Subwavelength imaging, wireless power transfer, antenna shrinking, and many more applications at various frequency bands have all made use of the new features of metamaterials. Wireless energy transfer and RF antenna minimization are the two fields that are most pertinent to coil design for MC. First, magnetic coils are mostly used in wireless energy transmission that employs the same antenna as MC. In [54], the metamaterial boost in energy transfer was proven using theoretical analysis. Second, electrically small antennas with high radiation efficiency are made possible by antenna-shrinking technologies, which can be used in MC to create small coils that increase the range of their communications. In [55, 56], an electrical dipole antenna encased by a metamaterial shell was examined in the deep subwavelength range. Greatly amplifying the far-field propagating waves and the power radiated by the electrical dipole is a promising method to increase the radiation efficiency of electrically tiny antennas.

We see great potential in using metamaterials to enhance the magnetic fields by radiating more power into the far field and ultimately achieve both extended medium penetration performance and a practical communication range. The main issue with MC is the rapid fall-off rate in the near field. The MI coil antenna is encased in a metamaterial-enabled resonant shell. Figure 4.2 serves as an example of the suggested metamaterial structure. The metamaterial shell's center contains the coil. Using such

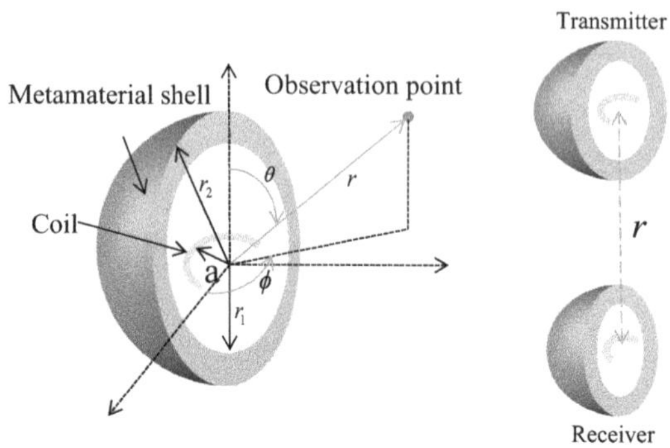

Figure 4.2 Illustration of the M^2I communication between two M^2I transceivers (MI coil enclosed in a metamaterial spherical shell)

a spherical shell can improve magnetic fields in any direction because coils produce spherical waves.

We first theoretically justify our assumption before demonstrating how this strategy might extend the range of MC. Designing and putting M^2I communications into practice presents three main difficulties: creating metamaterial spherical shell; accurately managing the effective permeability and thickness; and ensuring enhancement in both the near field and the far field. We then redesign the M^2I communication circuit model and present an EM analysis of the metamaterial shell.

4.2.2 Modeling and Analysis of M^2I Communications

Since it is unclear how to build the metamaterial, a theoretical model is helpful. Starting with an ideal analysis, we assume that the best possible metamaterial is accessible. In order to quickly expand the pathloss model to the M^2I waveguide, we first create it for point-to-point M^2I. Both parameters are applicable to the consideration of the ideal metamaterial shell configuration. In the analysis that follows, the underground soil with certain conductivity is regarded as the lossy medium, if no explicit remark is made.

By encasing the original MI coils in metamaterial spherical shells, the MC in the M^2I system is improved. By allowing waves to pass through the metamaterial layer, the near-field EM components can be altered and possibly improved. The corresponding circuit concepts are still applicable to M^2I but require significant revision. In particular, the properties of both the self-inductance L and mutual inductance M can be significantly altered by the metamaterial sphere. On the one hand, a significant increase in the mutual inductance in M^2I is anticipated. More magnetic flux passes between the transmitter and reception coils, increasing the mutual inductance because the metamaterial sphere's design goal is to enhance the near-field components of EM waves. On the other hand, even though the self-inductance L is a real number, it transforms into a frequency-dependent complex number in M^2I, i.e., $L = L_r - jL_i$, where L_r and L_i are the real positive values. The frequency-dependent resistance in the M^2I antenna is formed by the imaginary component of L (i.e., L_i), which has two distinct sources in

M^{2}I: the metamaterials and the complex surroundings. First, substantial eddy currents are created on the metallic components on the resonant metamaterial sphere, which is located very close to the coil itself. The primary magnetic field is opposed by a secondary magnetic field created by the eddy current, which lowers the coil's current and correspondingly raises its impedance. Second, because of the produced eddy current, the lossy medium also contributes to the imaginary portion of the self-inductance L. The updated impedance with imaginary inductance is $j\omega(L_r - jL_i) = j\omega L_r + \omega L_i$ since the impedance of an inductor is $Z_L = j\omega L$, where ω is the angular frequency. In order to obtain the magnetic resonance, the compensation capacitor changes to $C = \frac{1}{\omega_0^2 L_r}$.

Through the use of a capacitor, reactance is reduced at both the transmitter and the receiver. The load R_l is matched to the coil resistance R_c and the extra loss ωL_i, meaning that $R_l = R_c + \omega L_i$. Unlike EM wave-based wireless systems, M^{2}I's transmitter and receiver are tightly connected to one another, enabling an integrated transmitter–receiver system to perform impudence matching. Or, to put it another way, the transmitter is the source in the receiver, while the receiver is part of the loads in the transmitter.

We pay particular attention to researching the path loss in M^{2}I because the channel path loss largely determines the range of a wireless communication system. It should be noted that the pathloss analysis can also be used to obtain other crucial M^{2}I parameters, such as the channel bandwidth and channel capacity. Based on the equivalent circuit model and the prior considerations, the general pathloss formula in the M^{2}I channel between two transceivers (point-to-point communication) can be stated as follows:

$$\frac{P_r}{P_t} \approx \frac{\omega^2|M|^2 R_l/(R_l + R_c + \omega L_i)}{(R_c + \omega L_i)(R_c + R_l + \omega L_i) + \omega^2|M|^2}; \tag{4.9a}$$

$$\mathcal{L}_{p2p} = -10\log\left\{\frac{P_r}{P_t}\right\} \approx -20\log\frac{\omega|M|}{2(R_c + \omega L_i)}, \tag{4.9b}$$

where R_c is the coil's resistance, L_r is the coil's real self-inductance, C is the compensation capacitor used to tune the circuit to be resonant, R_l is the receiver's load, V_g is the source's voltage, and M is the mutual inductance between two adjacent coils.

The absence of impedance in the ideal power source is consistent with the ideal nature of the source in the following Comsol finite element method (FEM) simulations. The compensation capacitor $C = \frac{1}{\omega_0^2 L_r}$ is used to adjust the true inductance and produce circuit resonance, where ω_0 is the coil's resonant frequency. It should be emphasized that since there is no significant metamaterial amplification for original MI, the imaginary part of self-inductance is equal to zero. The load resistance R_l is designed to maximize (4.9a), which equals $R_c + \omega_0 L_i$. Equation (4.9a) can be approximated by (4.9b) at the resonant frequency in order to better understand the mechanics. ωM must be substantially smaller than $R_c + \omega L_i$ as a prerequisite. Given that L_i is quite large as a result of the resonance of the metamaterial shell, this approximation is useful. Furthermore, the mutual inductance is quite modest because we consider loose coupling for long-distance communication. As a result, the precondition is satisfied because ωM is substantially lower than R_c.

Similar to this, the M^{2}I waveguide is created by placing a metamaterial sphere on each coil in a magnetic waveguide (including the transmitter, receiver, and relays). Assume that the waveguide has $n - 2$ relaying coils. The transmitter coil is the first, while the receiving coil is the last. Each relay is thoughtfully positioned along a line with an interval of d. Using the corresponding circuit for M^{2}I waveguide, the generic route loss formula can be roughly written as follows:

$$\mathcal{L}_{wg} \approx 20(1-n)\log\left(\frac{\omega|M|}{R_c+\omega L_i}\right). \tag{4.10}$$

The path loss in the M^{2}I point-to-point communication and the M^{2}I waveguide, according to (4.9b) and (4.10), are the functions of the mutual inductance M, coil resistance R_c, and the frequency-dependent resistance L_i brought by the metamaterials and the propagation medium. Only M and L_i can be changed by the creation of the metamaterial sphere because R_c is a constant value. According to (4.9b) and (4.10), a basic approach to reducing path loss and expanding the communication range in M^{2}I is to increase mutual inductance M and decrease frequency-dependent resistance L_i. Due to the metamaterial shell's ability to concurrently magnify M and L_i, such a method is difficult to implement. We risk losing the gain of M if we reduce L_i. A new metric for M^{2}I communication, i.e., the inductance gain (IG) to characterize the advantage from the metamaterial shell, can be defined using the proportion $\frac{\omega|M|}{R_c+\omega L_i}$ to examine this tradeoff. The IG is represented mathematically as

$$\mathcal{G}_M = \frac{R_c^0|M^{\text{meta}}|}{(R_c^{\text{meta}}+\omega L_i^{\text{meta}})|M^0|}, \tag{4.11}$$

where M^{meta} and L_i^{meta} are the inductance enhanced by the metamaterial shell and M^0 is the mutual inductance without using the metamaterial shell. Keep in mind that L_i is comparatively minor and can be disregarded here without metamaterial (unless in a high conductive medium, such as seawater and under ocean environments).

4.2.3 Modeling the Metamaterial-Enhanced EM Fields in M^{2}I

According to the general framework of M^{2}I and the metamaterial enhancement technique covered in Section 4.2.1, the mutual inductance M and the complex self-inductance $L_r - jL_i$ play significant roles in M^{2}I communications. In this subsection, we analyze and model the metamaterial-manipulated EM field surrounding both the M^{2}I transmitter and the receiver in order to quantitatively characterize the impact of the metamaterial sphere on the mutual and self-inductance. The field model is then validated by using the FEM simulations.

It should be emphasized that in this subsection we concentrate on the M^{2}I point-to-point connection. Based on the examination of point-to-point communication, it is simple to determine the performance of the M^{2}I waveguide. In addition, the system's performance can be impacted by the coil orientation inside a metamaterial shell, particularly when two coils are perpendicular to one another. The coil arrays discussed in the chapter before, like the tri-axis coil, can be used to alleviate this issue. In M^2I, it is simple to put the tri-axis coil inside the metamaterial sphere. By simply combining the fields from the three coils, the performance is likewise simple to model. In the study what follows, we only take coils that are placed coaxially.

EM Fields around the M^2I Antenna

Take into account the M^2I point-to-point connection between an M^2I transmitter and an M^2I receiver as displayed in Fig. 4.2. In the middle of the metamaterial shell is the coil. The first and the third layers are the areas inside and outside the shell, respectively; the second layer is the actual shell. Standing waves are present in the first and second layers, while moving waves are present in the third layer. Therefore, the solution in the first two layers is built using spherical Bessel functions of the first kind and spherical Neumann functions. Only spherical Bessel functions of the first kind are employed in the first layer due to the singularity of the spherical Neumann function. The third layer makes use of the spherical Hankel function. For the ith layer, the wavenumber $k_i = \sqrt{\omega^2 \mu_i \epsilon_i}$, where $\epsilon_i = \epsilon_0 \epsilon_{ri} - j\frac{\sigma}{\omega}$, ϵ_0 is the vacuum permittivity, ϵ_{ri} is the relative permittivity, $\mu_i = \mu_0 \mu_{ri}$, μ_0 is the vacuum permeability, and μ_{ri} is the relative permeability. Also, a is the antenna radius, I_0 is the antenna current, r is the distance from the origin, η is the wave impedance, and **h** and **e** stand for the magnetic field and the electric field, respectively. The time dependence $e^{j\omega t}$ is assumed. In a lossy medium, the conductivity σ has a significant impact on the M^2I communication.

We first build the general solution to the wave equations in spherical coordination in accordance with classical EM theory. The whole solution can then be obtained depending on boundary conditions. Be aware that the antenna only emits TE_{01} since the radius of the magnetic dipole and the radius of its wire are significantly lower than the wavelength. We just need to think about the first-order mode because the metamaterial shell is significantly smaller than the wavelength. In order to express the unidentified magnetic fields in each layer, we have

$$\text{1st layer}: \begin{cases} \mathbf{h}_{r1} = \frac{-2j\cos\theta}{\omega\mu_1 r}\alpha_1 j_1(k_1 r)\hat{r}, \\ \mathbf{h}_{\theta 1} = \frac{j\sin\theta}{\omega\mu_1 r}\alpha_1 \left[j_1(k_1 r) + k_1 r j_1{}'(k_1 r)\right]\hat{\theta}; \end{cases} \tag{4.12a}$$

$$\text{2nd layer}: \begin{cases} \mathbf{h}_{r2} = \frac{-2j\cos\theta}{\omega\mu_2 r}\left[\alpha_2 j_1(k_2 r) + \alpha_3 y_1(k_2 r)\right]\hat{r}, \\ \mathbf{h}_{\theta 2} = \frac{j\sin\theta}{\omega\mu_2 r}\{\alpha_2\left[j_1(k_2 r) + k_2 r j_1{}'(k_2 r)\right] \\ \quad +\alpha_3\left[y_1(k_2 r) + k_2 r y_1{}'(k_2 r)\right]\}\hat{\theta}; \end{cases} \tag{4.12b}$$

$$\text{3rd layer}: \begin{cases} \mathbf{h}_{r3} = \frac{-2j\cos\theta}{\omega\mu_3 r}\alpha_4 h_1^{(2)}(k_3 r)\hat{r}, \\ \mathbf{h}_{\theta 3} = \frac{j\sin\theta}{\omega\mu_3 r}\alpha_4\left[h_1^{(2)}(k_3 r) + k_3 r h_1^{(2)\prime}(k_3 r)\right]\hat{\theta}, \end{cases} \tag{4.12c}$$

where α_i is the unknown coefficient; $j_1(kr)$ is the spherical Bessel function of the first kind of order 1, and $y_1(kr)$ is the spherical Neumann function of order 1, $h_1^{(2)}(kr)$ is the spherical Hankel function of the second kind of order 1, and the prime symbol denotes derivative.

The tangential component of the magnetic field (**h**) and the normal component of the magnetic flux (**B**) ought to be continuous at the boundary, according to Maxwell's equations. The unknown coefficients can then be obtained by rearranging (4.12) and adding the excitation source,

$$\mathbf{A}_t = \mathbf{S}_{\text{meta}}^{-1}\Psi_t, \tag{4.13}$$

where $\mathbf{A}_t^t = [\alpha_1, \alpha_2, \alpha_3, \alpha_4]$, $\mathbf{S}_{\text{meta}}$ is a coefficient matrix, and Ψ_t is the excitation vector. After solving (4.13), by substituting the unknown coefficients α_1, α_2, α_3, and α_4 into (4.12), the intensity of the magnetic field around the M^{2}I transmitter (outside the metamaterial sphere), i.e., $\mathbf{h}_{r3}$ and $\mathbf{h}_{\theta 3}$ in (4.12c), can be derived. The magnetic field intensity around the M^{2}I coil (inside the metamaterial sphere), i.e., $\mathbf{h}_{r1}$ and $\mathbf{h}_{\theta 1}$ in (4.12a), can also be derived.

The magnetic field in each layer of the receiver can be described by the same equation as the transmitter, using the (4.12). The third layer's magnetic field is different since it is the receiving shell's diffused magnetic field. Here, we can use the transformation of spherical coordinates to convert the coordinates from one spherical coordinate system to another. The incident magnetic fields can then be solved.

Validation Using FEM Simulations

In this subsection, we verify the developed field model via FEM simulation in Comsol Multiphysics; more thorough analysis can be found in [51]. The following settings have been made to the system. The metamaterial sphere is designed to have an overall diameter of 10 cm, or $r_2 = 0.05$ m. Many wireless devices can accommodate an antenna of this size. The operational frequency is set at 10 MHz, which is the same as other MI communication systems. Use is made of a single negative metamaterial layer. Without losing generality, we set the metamaterial's relative permeability to be -1, or $\mu_2 = -\mu_0$. With a relative permeability of $\mu_3 = \mu_0$, permittivity of $\epsilon_3 = 2\epsilon_0$, and conductivity of 2 mS/m, underground soil is considered to be the transmission medium outside the sphere.

By determining the ideal metamaterial layer thickness, or r_1, and the permeability of the infilling inside the sphere, or μ_1, the highest IG can be attained. We directly utilize the ideal values $r_1 = 0.025$ m and $\mu_1 = 5\mu_0$ because we simply need to validate the field model developed in this subsection. Furthermore, the MI coil a is required to be as big as it can be. We could theoretically set $a = r_1$. However, as a gets closer to $a = r_1$, it will have a significant impact on the border. We have used $a = 0.015$ m set. The radius of the coil without the metamaterial shell is set to 5 cm (the same as r_2) for fair comparison. It should be noticed that the aforementioned metamaterial sphere parameters are applicable.

It should be mentioned that metamaterial also takes into account the intrinsic loss impact. We think of μ_2 as having two different levels of losses, high loss and low loss. The corresponding parameters are: high loss $\mu_2 = (-1 - 0.05j)\mu_0$ and low loss $\mu_2 = (-1 - 0.005j)\mu_0$. The simulation is performed using the Comsol AC/DC module,

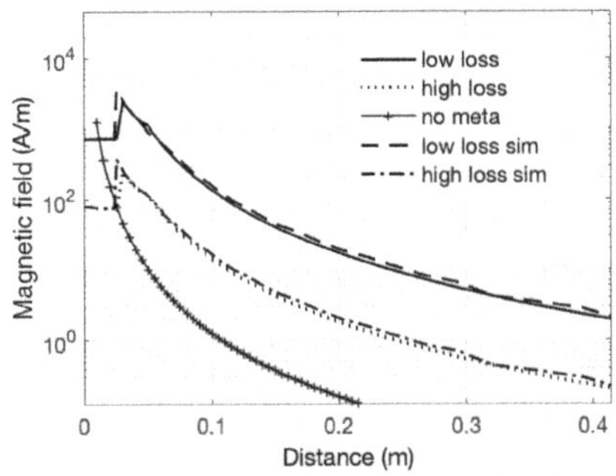

Figure 4.3 Magnetic field intensity around the transmitting coil in the soil medium (The center of the transmitting coil is located at 0 m, extending toward the receiver from 0 to 0.2 m)

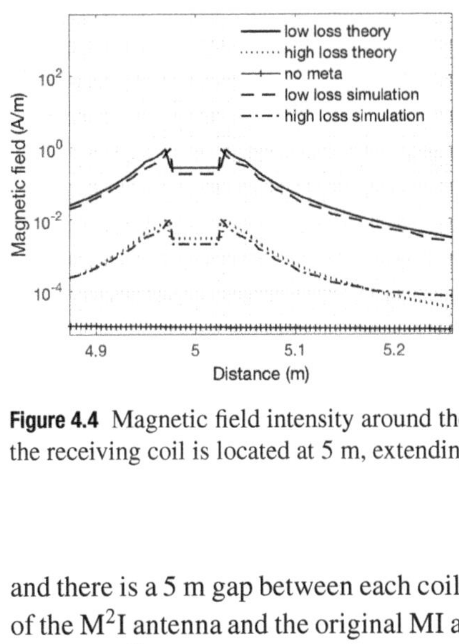

Figure 4.4 Magnetic field intensity around the receiving coil in the soil medium (The center of the receiving coil is located at 5 m, extending from 4.9 to 5.15 m)

and there is a 5 m gap between each coil. Instead of the same antenna current, we think of the M^{2}I antenna and the original MI antenna as having the same transmission power P_t. The M^{2}I coil, as previously mentioned, includes additional frequency-dependent resistance from the self-inductance, which can use a lot of power. The extra resistance (ωL_i) has a significant impact on the input impedance for the four cases under consideration. For all the three cases, we set the transmission power P_t in (4.9b) to 1 W in order to ensure an equal comparison.

The magnetic field strength produced from the theoretical field model matches well with the findings of the FEM simulation at both the transmitter and the receiver sides, as illustrated in Figs. 4.3 and 4.4. On the one hand, we see that the magnetic field intensity may be raised by more than one order of magnitude in comparison to the original MC system by utilizing the metamaterial shell with optimal settings. On the other hand, the gain provided by the metamaterial sphere can be significantly reduced by the metamaterial loss. Additionally, note that the transmitter side has a smaller gain than the reception side. The magnetic field is amplified once more by the receiver's metamaterial shell, which is the reason of the improved magnetic field strength.

4.2.4 Analytical M²I Channel Model

The self-inductance and mutual inductance as well as other channel parameters in the M²I communication can be computed using the field model created in the previous part. But the field model necessitates challenging calculations, such as the inverse of the Bessel function matrix. As a result, the model is restricted to numerical data and is unable to offer an analytical understanding of the mechanism underlying metamaterial enhancement, let alone system optimization. In order to achieve this, we create the analytical M²I channel model with explicit and understandable formulations for bandwidth, capacity, self-inductance, mutual inductance, and path loss. The resonance condition and the ideal configuration for M²I communications are then examined by using the analytical model as a foundation. Detailed derivation of this part can be found in [51].

Deriving Explicit Expressions for an Analytical Channel Model

On the basis of the constructed field model, we begin the investigation by computing the M²I self-inductance and mutual inductance. Due to the metamaterial shell's impact, the self-inductance is divided into two components: the initial inductance produced by the coil and the additional inductance provided by the metamaterial improvement:

$$L = \frac{\Phi_1}{I_0} = \frac{1}{I_0} \iint_S \mathbf{B} \cdot \hat{n} \, dS \simeq L_0 + \frac{4\pi\alpha_1}{j\omega k_1 I_0}\left[1 - \frac{\sin(k_1 a)}{k_1 a}\right], \tag{4.14}$$

where Φ_1 is the magnetic flux through the transmit coil, S is the area of the coil, $\hat{n}$ is the orientation of the coil, and L_0 is the coil's self-inductance without the shell, which can be approximated by $L_0 = \mu_1 a[\ln(\frac{8a}{r_w}) - 2]$, where r_w is the wire radius. Similarly, the mutual inductance is the magnetic flux through the receive coil over the current in the transmit coil, which can be expressed as

$$M \simeq \frac{4\pi\beta_1}{j\omega k_1 I_0}\left[1 - \frac{\sin(k_1 a)}{k_1 a}\right]. \tag{4.15}$$

Keep in mind that because the reradiated field from the receiver coil is bidirectional, it is taken into account in this mutual inductance.

The key coefficients that determine the self-inductance and mutual inductance are α_1 and β_1. To derive β_1, α_4 is also needed to be calculated. Those coefficients need to be derived through (4.13), which require the calculation of the inverse of a matrix consisting of different types of Bessel functions. To derive a tractable channel model, such functions need to be simplified. Fortunately, since the shell and the antenna are electrically small, i.e. $k_3 r_2 \ll 1$, and L and M can be simplified as

$$\widetilde{L} \approx L_0 + \frac{\pi\rho_1 a^4 \mu_1 \left[r_2^3(\mu_1 - \mu_2)(\mu_2 + 2\mu_3) + r_1^3(\mu_2 - \mu_3)(2\mu_1 + \mu_2)\right]}{18\rho_2(\rho_{3i} + \rho_{3r} j)^2 r_1^4 r_2^4 \mu_2 \mu_3 \det(\widetilde{\mathbf{S}}_{\text{meta}})}, \tag{4.16}$$

$$\widetilde{M} = \frac{-j\pi a^4 \rho_1^2 \mu_1^2 (\rho_{3i} - \rho_{3r} j)^2}{2\ \rho_2^2 r_1^2 r_2^2 \mu_3 (\rho_{3i}^2 + \rho_{3r}^2)^2} \frac{\hbar(r)}{[\det(\widetilde{\mathbf{S}}_{\text{meta}})]^2}, \tag{4.17}$$

where $\rho_1 = k_1$, $\rho_2 = -jk_2$, $\rho_{3r} = \Re(k_3)$, and $\rho_{3i} = -\Im(k_3)$, and all $\{\rho_x\}$ are real positive numbers, and $\hbar(r)$ is a function of distance that is determined by the antenna pattern. Here $\det(\widetilde{\mathbf{S}}_{\text{meta}})$ is the approximation of $\det(\mathbf{S}_{\text{meta}})$, which is

$$\begin{aligned}\det(\mathbf{S}_{\text{meta}}) &\approx \det(\widetilde{\mathbf{S}}_{\text{meta}}) + o(\frac{1}{\bar{k}\bar{r}}) \\ &= \frac{\rho_1 \left[2r_1^3(\mu_1 - \mu_2)(\mu_3 - \mu_2) - r_2^3(2\mu_2 + \mu_1)(2\mu_3 + \mu_2)\right]}{9\rho_2(\rho_{3i} + \rho_{3r} j)^2 r_1 r_2^4 \mu_2 \mu_3} + o\left(\frac{1}{\bar{k}\bar{r}}\right),\end{aligned} \tag{4.18}$$

where $\bar{k}$ is the asymptotic approximation of all the wavenumbers and $\bar{r}$ is the asymptotic approximation of all the radii.

Optimal Configuration of M^{2}I Communications

The transmitter and the receiver are connected by the mutual inductance M. According to (4.17), M can be maximized if $\det(\widetilde{\mathbf{S}}_{\text{meta}})$ is zero (i.e., $\det(\mathbf{S}_{\text{meta}})$ is very small). There exists an optimal metamaterial sphere thickness r_1 that can greatly reduce the value of $\det(\mathbf{S}_{\text{meta}})$. As a result, the mutual inductance M can be significantly increased. Then the magnetic field intensity around both the M^{2}I transmitter and the receiver can be dramatically enhanced. The condition to achieve such an enhanced peak is to find r_1 that makes $\det(\widetilde{\mathbf{S}}_{\text{meta}}) = 0$. The solution to $\det(\widetilde{\mathbf{S}}_{\text{meta}}) = 0$ can be developed as

$$\frac{r_1}{r_2} = \sqrt[3]{\frac{(2\mu_3 + \mu_2)(2\mu_2 + \mu_1)}{2(\mu_2 - \mu_3)(\mu_2 - \mu_1)}}. \tag{4.19}$$

The metamaterial sphere reaches the metamaterial resonance if it fulfills (4.19). Since $r_1 < r_2$ prevents such resonance from occurring if μ_2 is positive, the use of metamaterials as the sphere's second layer is required. Additionally, μ_1 can be increased to more than $100\mu_0$ if a ferrite core is used for the coil antenna to enhance performance. We can still locate the resonance configuration by modifying r_1. Figure 4.5 displays an illustration of resistance in a resonance configuration.

The metamaterial antenna designs of [55] and [57] discovered this resonance condition, suggesting that the resonance is the best operating mode because it amplifies the radiated power in the distant field to the greatest extent. The performance of the antenna can be considerably harmed by any departure from resonance. The radiated power is less significant in the near-field than it is in far-field communications, which is where the M^{2}I communications rely. In lossy media, the resonance also increases the frequency-dependent resistance ωL_i in addition to the mutual inductance M. As a result, a thorough reexamination of the role of metamaterial resonance in M^{2}I communications is required.

The frequency-dependent resistance ωL_i comes from the imaginary part of the self-inductance L. Hence, we investigate the effect of metamaterial resonance on the

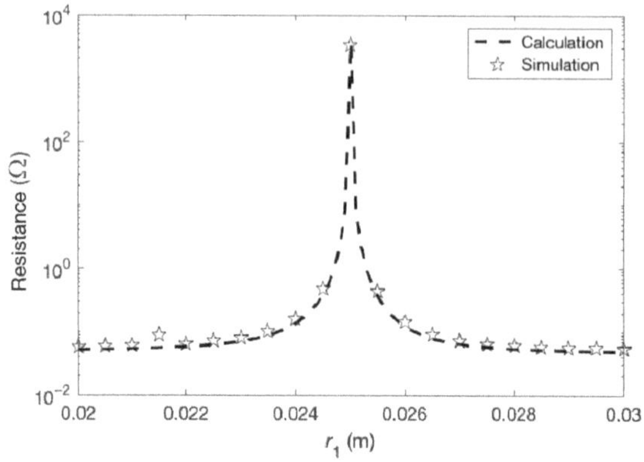

Figure 4.5 Resistance versus r_1

self-inductance L given in (4.16). Under the resonant condition in (4.19), $\tilde{L}$ can be updated as

$$\tilde{L}^{\mathrm{rc}} \approx L_0 + \frac{\rho_1 \pi a^4 \mu_1 \left[r_2^3(\mu_1 - \mu_2)(\mu_2 + 2\mu_3) + r_1^3(\mu_2 - \mu_3)(2\mu_1 + \mu_2) \right]}{18\rho_2(\rho_{3i} + \rho_{3r} j)^2 r_1^4 r_2^4 \mu_2 \mu_3 \cdot o(\frac{1}{k\bar{r}})}. \tag{4.20}$$

In (4.20), the absolute value of the second term is maximized since $o(\frac{1}{k\bar{r}})$ is the minimum value of $\det(\mathbf{S}_{\mathrm{meta}})$. If the transmission medium is lossless, such as the air medium in most existing works, the second term in (4.20) is real since the wavenumber of the medium k_3 is real (i.e., $\rho_{3i} = 0$). Therefore, the self-inductance L is real, which can be compensated by the capacitor. As a result, the M^2I system experiences no additional loss even when the self-inductance is maximum. However, the wavenumber k_3 becomes complicated if the lossy medium taken into account in this thesis. Consequently, the imaginary part of L (i.e., the frequency-dependent resistance) in (4.20) is also maximized when it is resonant, which causes a significant loss in M^2I.

Figure 4.5 shows the total resistance ($R_{\mathrm{coil}}^{\mathrm{total}} = R_c + \omega L_i$) of an M^2I coil in the soil medium as a function of the sphere thickness r_1, based on both the developed model and FEM simulations. As predicted, the coil resistance is extremely large when the sphere is resonant ($r_1 = 0.025$ m). Hence, in M^2I, the resonance condition amplifies both the mutual inductance M and ωL_i. As the sphere thickness r_1 moves away from the resonant condition, L_i approximates to 0. As a result, the frequency-dependent resistance disappears and only the coil wire resistance is left, i.e., $R_{\mathrm{coil}}^{\mathrm{total}} \approx R_c$.

According to (4.11), the IG $\mathcal{G}_M$ between the M^2I transceivers is, in fact, determined by the ratio M/L_i. The effect of resonance on $\mathcal{G}_M$ is not clear yet since both M and L_i are maximized at the resonance point. However, according to (4.14) and (4.15), M is inversely propositional to $[\det(\widetilde{\mathbf{S}})]^2$, while L is inversely propositional to $\det(\widetilde{\mathbf{S}})$. Given that the resonance condition is actually $\det(\widetilde{\mathbf{S}}) = 0$, M is greatly amplified when resonant compared to L_i. Therefore, we may say that the metamaterial resonance still represents M^2I's ideal operational state. However, when r_1 deviates from the ideal value, the system performance does not degrade as quickly as existing metamaterial antennas due to the similar resonance effect of the frequency-dependent resistance (which results in loss). As a result, the M^{2}I system is less sensitive to size variances, which is advantageous for fabricating real devices.

4.2.5 M^{2}I Waveguide

When several M^{2}I devices are arranged in a straight line and the distance between them is narrow enough, the M^{2}I waveguide can be created. For instance, multiple M^{2}I sensor nodes can be densely placed in applications of wireless sensor networks. Multiple M^{2}I nodes exist between the M^{2}I transmitter and receiver, forming an M^{2}I waveguide along the transmission line.

The fundamental idea behind M^{2}I waveguide enhancement is that strong connections between individual coils have resulted from a greatly improved coupling between them. We can think about it in a different way. The signal propagation loss along the transmission line is minimal if the transmitter and the receiver are connected by a wired transmission line. Since the connection between the transmitter and the receiver is now weaker, the signal propagation loss increases if the original magnetic waveguide serves as the transmission line. A stronger coupling is possible with the M^{2}I waveguide. Although it is not strong enough to be referred to as a wire, it is nevertheless far more powerful than the original magnetic waveguide, which reduces the signal propagation loss.

4.3 Antenna Design for Practical M^{2}I Communication

The metamaterial discussed in earlier parts is homogeneous and isotropic in nature, or ideal. It takes work to design and make this form of metamaterial. Metamaterials have been developed, and they are now used in optical and cell phone antennas. There aren't many uses of metamaterial in the HF or lower frequency regions. The primary antenna and metamaterial design goals for M^{2}I communication are initially presented in this section. Then, in order to achieve the design goals, we introduce the spherical coil array-enabled metamaterial shell. Finally, we demonstrate the proposed structure's manufacturing and performance testing. More details can be found in [58].

4.3.1 Metamaterial Modeling and Analysis

When a mixture contains particles that are significantly smaller than its wavelength, it can be said to be an effective homogeneous medium, and its constitutive properties, including its effective permeability, permittivity, and conductivity, can be calculated analytically [59]. As a result, a lot of atoms with sizes that are considerably smaller than the wavelength are required to create the metamaterial. These atoms can alter the nature of how waves propagate to show unusual phenomena all at once. As a result, how we build the atoms will define the characteristics of this form of artificial material. The following section will start with an illustration of how a coil might alter

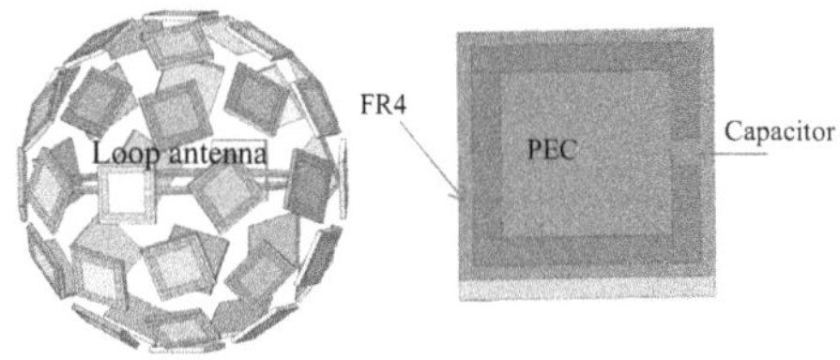

Figure 4.6 Spherical coil array enclosed loop antenna and a coil unit on the shell

the propagation of a magnetic field. The coil array is then cooperatively designed, and its equivalence to a metamaterial shell is demonstrated.

Consider a source that emits a magnetic field $H_s\hat{s}$ that can cause current I_s to flow via a coil. Here, the magnetic field's direction is represented by the unit vector $\hat{s}$. The magnetic flux can be written as $B_m = \mu_0 H_s\hat{s}$ if we assume that there is no coil, where μ_0 is the permeability of vacuum. The coil will reradiate the magnetic field $H_c\hat{c}$, where $\hat{c}$ is likewise a unit vector designating the direction of the reradiated magnetic field, if the coil leaves due to the induced current I_s. Consequently, the magnetic flux at the portion of the second coil can be updated as

$$B_m = \mu_0 H_s\hat{s} + \mu_0 H_c\hat{c} = \mu_0\left(1 + \frac{H_c\hat{c}}{H_s\hat{s}}\right)H_s\hat{s} = \mu_{\text{eff}} H_s\hat{s}, \tag{4.21}$$

where μ_{eff} is the effective permeability at point o.

In light of (4.21), it is possible to alter the effective permeability as long as the reradiated magnetic field $H_c\hat{c}$ may be changed. By altering the capacitor in the coil, this can be accomplished. The reradiated magnetic fields can be highly varied as a result of the coil's impedance shift, which can lead to various field propagation patterns.

Figure 4.6 displays the metamaterial's geometrical structure. The outside radius of the spherical shell was fixed at 0.05 m, just as we discussed in the ideal model. We evenly divide the sphere's surface into 80 triangles with 42 vertices using the Pentakis icosidodecahedron. A coil has its center at each vertex, and the orientation of the coil is the radial direction from the shell's center to the vertex. The coil array is created using 42 identical coils in this fashion. The simulation models of a printed circuit board (PCB) unit and a related equivalent circuit are shown in Fig. 4.6. The coils are made of copper with capacitors to tune them. The coil inductance, resistance, and compensating capacitance are each denoted in the equivalent circuit by the letters L_c, R_c, and C. Due to its size being significantly less than the wavelength, the square coil used here performs similar to a circular coil. Here, we demonstrate that this structure is equal to the ideal metamaterial shell and can achieve negative permeability.

First, a homogeneous layer is thought of for the spherical coil array. We determine the effective permeability of this layer using the effective medium theory [60] of magnetism. As illustrated in Fig. 4.7, the radiation source is the same as it is in the ideal form, a magnetic loop antenna, and all of the coils on the shell are passive. Kirchhoff's voltage law allows us to determine

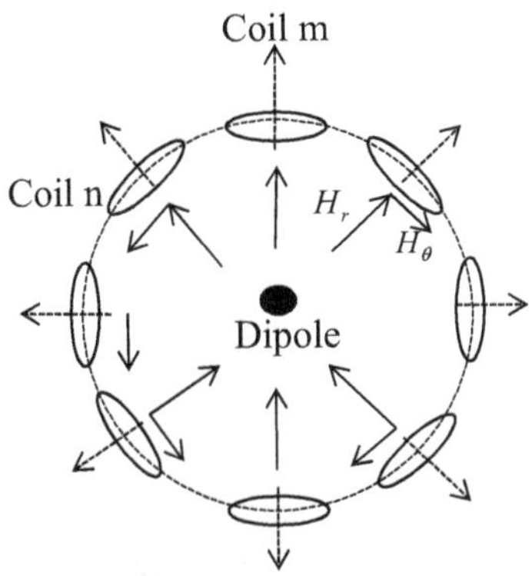

Figure 4.7 Illustration of coil orientation and the direction of the magnetic field radiated by a loop antenna

$$\begin{bmatrix} Z_1 & j\omega M_{12} & \cdots & j\omega M_{1N} \\ j\omega M_{21} & Z_2 & \cdots & j\omega M_{2N} \\ \vdots & \vdots & \vdots & \vdots \\ j\omega M_{N1} & j\omega M_{N2} & \cdots & Z_N \end{bmatrix} \cdot \begin{bmatrix} I_1 \\ I_2 \\ \vdots \\ I_N \end{bmatrix} = \begin{bmatrix} V_1 \\ V_2 \\ \vdots \\ V_N \end{bmatrix}, \tag{4.22}$$

where N is the number of coils on the shell, $Z_n = R_c + j\omega L_c + 1/j\omega C$, M_{mn} is the mutual inductance between coil m and coil n, I_n is the induced current in coil n, and V_n is the voltage. The magnetic fields radiated by a coil or loop antenna include the radial direction H_r and the polar direction H_θ.

Figure 4.7 shows the direction of H_r and H_t. Only H_r can induce currents in the coils on the shell because the coil orientation is parallel to the H_r's direction and perpendicular to the H_t's direction. This shell can therefore be thought of as a metamaterial for H_r but not for H_t. A tri-axis coil or a cubic metamaterial, like that in [61, 62], can be used to resolve this. Here, we merely take into account H_r in order to generate concise analytical conclusions from which to extrapolate deeper insights. As a result, V_n in (4.22) can be written as

$$V_n = -j\omega\pi a^2 \mu_0 H_r \cos\theta_n, \tag{4.23}$$

where θ_n is the angle of H_r and the coil's orientation. According to Kirchhoff's voltage law, without loss of generality, in coil n, we can obtain

$$I_n(R_c + j\omega L_c - \tfrac{j}{\omega C}) + \textstyle\sum_{i=1, i\neq n}^{N} j\omega M_{in} I_i = -j\omega\pi a^2 \mu_0 H_r \cos\theta_n. \tag{4.24}$$

By looking at the equation above, we can see how a coil's induced voltage changes as a result of its neighbors. Here, we can assume that the metamaterials caused this change. The strength and propagation of the magnetic field have been altered due to the presence of a coil's array neighbor. We can therefore produce either positive or negative impacts as a result. In [58], the effective permeability is given as

$$\mu_{\text{eff}} = \left(1 - \frac{j\omega \sum_{i=1,i\neq n}^{N} M_{in}}{R_c + j\omega L_c + 1/j\omega C + \sum_{i=1,i\neq n}^{N} j\omega M_{in}}\right)\mu_0. \tag{4.25}$$

We can see from (4.25) that a negative permeability can be reached when the second term in the parenthesis is greater than 1. Furthermore, we can see from (4.25) that the effective permeability is solely governed by the mutual inductances of the coils and the lumped elements. The radiation source has no effect on it. To create a spherical negative permeability metamaterial, we have already solved the first two problems. Then, we demonstrate how to alter the effective permeability of the metamaterial as desired and identify the ideal conditional parameters.

4.3.2 Optimal Design of the M²I Antenna

The second term in the parenthesis in Eq. (4.25) must be maximized in light of (4.25) in order to produce a significant negative permeability. We pay particular attention to the real element of the effective permeability since the fictitious portion stands for loss. In order to distinguish the genuine part, Eq. (4.25) can be written as follows:

$$\mu_{\text{eff}} = \left[1 - \frac{\omega \sum_{i=1,i\neq n}^{N} M_{in} X}{R_c^2 + X^2} - j\frac{\omega \sum_{i=1,i\neq n}^{N} M_{in} R_c}{R_c^2 + X^2}\right]\mu_0, \tag{4.26}$$

where $X = \omega L_c - \frac{1}{\omega C} + \omega \sum_{i=1,i\neq n}^{N} M_{in}$. In order to obtain negative permeability, $F = \frac{\omega \sum_{i=1,i\neq n}^{N} M_{in} X}{R_c^2+X^2}$ need to be maximized to let it at least be larger than 1. Then, we find the derivative of F with regard to X, and the optimal condition is $X^2 = R_c^2$. Hence, X can be positive R_c or negative R_c. The term $\sum_{i=1,i\neq n}^{N} M_{in}$ is negative, and thus when $X = R_c$, a large positive $\Re(\mu_{\text{eff}})$ can be obtained, while when $X = -R_c$, a large negative $\Re(\mu_{\text{eff}})$ can be obtained, which is $1 + \frac{\omega \sum_{i=1,i\neq n}^{N} M_{in}}{2R_c}$. Since our objective is to design negative permeability metamaterial, $X = -R_c$ is the prerequisite, upon which we can obtain the resonance frequency:

$$\omega = \frac{-R_c \pm \sqrt{R_c^2 + 4\left(L_c + \sum_{i=1,i\neq n}^{N} M_{in}\right)/C}}{2\left(L_c + \sum_{i=1,i\neq n}^{N} M_{in}\right)}. \tag{4.27}$$

We observe that there is a meaningful positive solution and a nonsensical negative solution in (4.27). Because there is just one resonant frequency, the permeability is strongly negative. Noting that we may also easily adjust the capacitance C in addition to the frequency, By adjusting C, we can also accomplish the resonance if ω_0 is the intended operating frequency, and the ideal value is

$$C = \frac{1}{\omega_0\left(R_c + \omega_0 L_c + \omega_0 \sum_{i=1,i\neq n}^{N} M_{in}\right)}. \tag{4.28}$$

Because when the real portion is considerable, this also creates a large imaginary part, the above ideal condition can only predict the most negative permeability, which may not be the optimal condition of M²I communication. Negative permeability, however, can only be obtained in a very small band or capacitance range since it depends on the resonance of the coil unit. As a result, the aforementioned ideal situation can still serve as a decent approximation. We may see that R_c mostly determines the imaginary portion of the effective permeability by looking at (4.26).

In particular, if we ideally assume that R_c is 0, the imaginary component vanishes, and we may get any μ_{eff} we choose by altering X or, equivalently, changing C. Therefore, the coil's resistance R_c becomes the bottleneck and must be lowered to lessen the undesirable effects. We can aggressively widen the trace, utilize a lower frequency to lessen AC resistance, or shorten the trace to reduce R_c. However, these answers are either impractical or ineffective. Here, we demonstrate that since ω increases faster than R_c at lower MHz bands, increasing the operating frequency allows us to obtain more negative μ_{eff}. To put it another way, by using a higher frequency, we can increase the ratio $\frac{\omega \sum_{i=1, i \neq n}^{N} M_{in}}{2R_c}$, which is comparable to maintaining ω while reducing R_c.

4.4 Wireless Communication Performance Analysis

A model to produce negative effective permeability has been presented in Section 4.3.2. We may adjust the spherical coil array's properties, particularly the capacitance, using this model. Finding resonance parameters in this way allows us to improve the magnetic fields that are produced. In contrast to the parts that came before, which concentrated on the analysis and design of antennas, in this section, we pay attention to the performance of wireless communication, specifically the path loss.

4.4.1 Radiation Efficiency and Analysis

We use a similar methodology as [63], where the transmitting antennas receive the same input current of 1 A, and the magnetic field intensity radiated by the antennas is analytically calculated and numerically simulated in order to compare the radiation efficiency of M^2I communication and MI communication. Note that the MI loop antenna's radius is set as 0.05 m, which is also the M^2I shell's outer radius, compared to the M^2I loop antenna's radius of 0.04 m. Even though they have the same negative permeability, the performance of the ideal M^2I is not comparable to that of coil array-based M^2I antennas, because the inner radius of the ideal M^2I antenna in the ideal model is 0.025 m and the loop antenna's radius is 0.015 m. The coil array-based M^2I antenna, however, has essentially nearly zero thickness. In order to utilize the space to the full, we can set up the loop antenna significantly larger than the ideal M^2I antenna.

Magnetic Field Analysis

In M^2I communication, only the main loop antenna is actively stimulated; all other small coils on the shell are passive, as shown in Fig. 4.6. We utilize the induced voltage in the receiving coil as a metric to demonstrate the amplification because it is proportional to the strength of the magnetic field. The loop antenna is the only component of the MI antenna that emits magnetic fields. However, with the M^2I antenna, the magnetic field is the vector superposition of that radiated by the loop antenna inside the shell and the reradiated magnetic fields by the passive coils on the shell, which may be represented as

$$\hat{H} = \sum_{i=0}^{N} \left[H_r(I_i)\hat{r}_i + H_t(I_i)\hat{\theta}_i \right], \tag{4.29}$$

where the large loop antenna is denoted by $i = 0$. To find I_i, we can use (4.22) and (4.23).

Physical Principle of Enhanced M^{2}I Communication

The results of the theoretical analysis, which makes use of ideal EM theory, demonstrate the enhancement. We examine a spherical coil array in the actual design and demonstrate that it is equivalent to metamaterials, which, if we can achieve resonance, can also improve magnetic fields. Then, we demonstrate this strategy's operation by using a more intuitive understanding.

Only when the dominating component becomes zero, the radiated magnetic field can be strengthened, and a noticeably high mutual inductance can be attained, as covered in the sections before. In order to allow μ_2 to be negative, we must utilize metamaterial. Now, viewing the shell from a different angle, we demonstrate how each single coil functions to enable the shell become resonant.

We can alter the effective permeability by modifying the reradiated magnetic field using coils, as demonstrated in Section 4.3.1. To augment or cancel the magnetic field radiated by the loop antenna, we can modify the direction of the reradiated field by adjusting the capacitance. To further understand the physics, we assume that the small coil's reactance $X = \omega L_c - 1/(\omega C) + \omega \sum_{i=1, i \neq n}^{N} M_{in}$ can be either positive or negative and that its resistance R_c is indefinitely small or comparable to a no-loss metamaterial unit. Next, suppose that a small coil experiences a voltage $V = -j\omega\pi a^2 \mu_0 H_s = -jV_0$ as a result of the magnetic field H_s that the loop antenna emits. To distinguish between resonance and nonresonance as well as the positive and negative peaks, we now take three scenarios into account.

First, although the induced voltage is high, the induced current in the small coils is incredibly low when the small coil is far from resonance, i.e., X is drastically big. Since only the loop antenna in the middle is radiating, the impact of those tiny coils on the shell can be disregarded. Due to the fact that the M^{2}I's radiating loop antenna is enclosed in a smaller shell than the MI's antenna, it has no advantage over MI in this situation. What's more, M^{2}I is less efficient than MI. Because of this, the field intensity of M^{2}I is weaker than MI on both sides of the resonance.

Let's now examine the resonance region. The first peak is encouraging. Since the capacitance that can cause the reactance to disappear, C, is less at this point, $X < 0$. The current in each tiny coil, $V_0/|X|$, has the same direction as the current flowing through the radiating loop antenna when R_c is insignificant. As a result, the near-field power of the radiating loop antenna can be captured by the shell's coils and broadcast again. Because X is small and the small coil is close to resonance, a big current flow through the coil, making the phenomena intense. The reradiated field is important as a result.

Since C exceeds the capacitance that can cause the reactance to vanish at the negative peak, $X > 0$. The current in a tiny coil at this time is $-V_0/|X|$ and is flowing in the opposite direction to that of the radiating loop antenna's current. As a result, the loop antenna's magnetic field is canceled by the reradiated magnetic field. Similar to this, as X increases more, the induced current in the little coil decreases noticeably, lessening the undesirable effect.

In addition to the model-based knowledge of M^2I communication provided in the ideal model, we are able to provide a more logical justification of the enhancement mechanism of M^2I communication in this context from the viewpoint of realistic design and execution. As previously mentioned, the incredibly inefficient antenna, which results from the relatively small coil size compared to the signal wavelength, is the primary cause of the original MI's extremely constrained communication range. Around the initial ineffective MI antenna, the reactive power to real power ratio is roughly $(\frac{\lambda}{2\pi a})^3$ [64]. Reactive power, which only exists in the immediate region of the coil antenna and cannot be used for wireless communications if the transmission distance is quite great, makes up the majority of the power produced by the coil antenna.

The M^2I communication, on the other hand, can make use of such reactive power by turning a portion of it into real power that the MI communication can rely on. Reactive power specifically causes current to flow via the tiny coils on the metamaterial shell. The tiny coil then emits actual power and reactive power similarly to a loop antenna. The reactive power is converted to real power using a larger amount of tiny coils relative to the initial real power. This explains why the transmitting antenna's distant and near regions can both benefit from the metamaterial amplification of M^2I.

4.5 Implementation and Experimental Analysis

Using a 3D printed spherical frame and printed coils on a circuit board, we created the M^2I antenna in this section. The wireless channel between M^2I transceivers is measured in various situations, and the magnetic field enhancement is verified.

4.5.1 Antenna Implementation for M^2I Communication

The M^2I antenna consists of the 3D printed spherical frame and the metamaterial units.

3D Printed Spherical Frame

To support the coil-antenna array, a spherical frame is created using a 3D printing technology. A new disruptive technology called 3D printing allows us to create 3D objects directly from digital models without the use of special tools or fixturing. As a result, it is more adaptable and effective than conventional production procedures that rely on molding and machining. Fused deposition molding (FDM) is one of most common 3D printing technology that is advantageous in terms of material property, cost, and accessibility compared with other technologies. In this research, the FDM-based

Figure 4.8 Implementation

desktop printer (MakerBot Replicator 2X) has been applied to create the spherical form support framework for the coil-antenna array.

The first step in this procedure is to create a 3D digital model of the final product that represents its geometry and topology. This model is created using the commercial computer-aided design (CAD) program Creo Pro, from PTC (Boston, MA), as illustrated in Fig. 4.8. The digital model is divided into many two-dimensional (2D) layers with a thickness of a few micrometers by using specialized model pre-processing software (Makerware), and the layers are then rasterized using tool path planning. The 3D printer receives the path data after which it uses its X, Y, and Z axes to draw the model one layer at a time. Filament is fed into the printer, heated to the material's melting point, and then extruded through a nozzle with a small aperture. Layer by layer, the spherical frame is constructed.

Metamaterial Units and Coil-Antenna Array

As covered in the preceding sections, the M^2I antenna's metamaterial layer is effectively realized by a spherical array of tiny coils. In our design, the small coils are placed on the 3D printed frame to create the spherical coil array. The little coils and spherical frame are precisely of the same size as the simulation model from Section 5.1. Circuit boards with a thickness of 1.57 mm have the tiny coils printed on them. The copper is one ounce thick. The square PCB's edge is 18 mm long. The coil's outer edge measures 16 mm, and the trace's width is 2 mm. The positions of each coil match those in the simulation model. Figure 4.8 displays the finished spherical frame with metamaterial units (i.e., the little coils) that was 3D printed.

4.5.2 Wireless Coupling Enhancement

The magnetic field radiated by a magnetic dipole can be strengthened by the metamaterial shell, as covered in Section 4.4.1. As shown in (4.23), the voltage is proportional to the magnetic field, making the S_{21} parameter a good indicator to show the improvement of the wireless coupling. We use this parameter to support this conclusion because it is proportional to the ratio of the transmitted voltage to induced voltage.

We measure the S_{21} parameter with a metamaterial sphere (S_{21}^{w}) and without a metamaterial sphere (S_{21}^{wo}), using a pair of nonresonant magnetic loop antennas with a radius

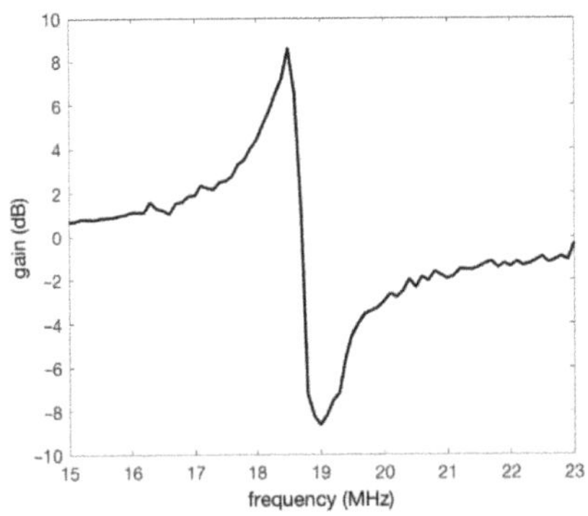

Figure 4.9 S_{21} parameter gain. The S_{21} of M^{2}I in dB scale minus the S_{21} of MI in dB scale

of 0.045 m, similar to the technique used in [61] to assess the performance of metamaterial. The S_{21} gain is defined as $S_{21}^{w} - S_{21}^{wo}$, where S_{21} parameters are in dB scale. The Agilent 8753E RF network analyzer (Agilent, Santa Clara, CA) measures the S_{21} parameter. The tiny coils on the sphere are tuned using 2700 pF capacitors.

A resonance is reached at 18 MHz, and the gain is approximately 8 dB, as illustrated in Fig. 4.9. The findings demonstrate that the resonant peak at 18 MHz and the null immediately following it exactly match our calculations and predictions for the radiation efficiency. It's important to remember that the M^{2}I implementation calls for the capacitors to have low tolerance and very similar tiny coils on the shell. Our capacitors have a tolerance of 1 %.

The small coils on the sphere will have slightly different resonance frequencies if the tolerance is larger, which will result in an uneven distribution of the sphere's negative permeability or, in the worst-case scenario, positive effective parameter in parts of the coils. Similar to this, the coils must be similar to prevent differences in self-inductance and resistance from also causing inhomogeneity on the sphere.

4.5.3 Wireless Channel Measurement

We employ USRP software (Ettus Research, a National Instruments Brand, Austin, TX) defined radio kits as the signal transmitting and receiving apparatuses to measure the path loss of the wireless channel between two M^{2}I transceivers. The USRP boards are coupled with the M^{2}I antennas. The USRP N210 mother board that we use is based on a Xilinx Spartan-3A DSP 3400 FPGA. It has a 400 MS/s dual DAC and a 100 MS/s dual ADC, and it is gigabit ethernet connected to the computer. Through connectors TxA/RxA and TxB/RxB, the daughter boards LFTX/LFRX can handle two separate antennas. This daughter board can produce and receive wireless signals in the range of 0–30 MHz, which falls inside the 18 MHz frequency band that we are interested in.

Based on these hardware devices, a computer's GNU software generates and analyzes the signal. A signal source block with a frequency of 100 kHz produces the signal, and an 18 MHz transmission frequency is used. The received time domain signal is transformed into a frequency domain signal at the receiving end using a fast Fourier transformation (FFT) block. A thorough investigation of the power loss and gain of every component the signal passes through is required to obtain the precise power value in dBm. The measurements were performed in various environments, and more details can be found in [58].

Part II

Theoretical Basis

5 MI Network Connectivity

5.1 Wireless Network Connectivity

5.1.1 Connectivity of Graphs

Let's start understanding the connectivity problem with a basic graph $G = (V,E)$, where V represents a set of vertices and $E \subseteq \{(x,y)|(x,y) \in V\}$ is a set of edges. Two nodes are neighbors if they have an edge in E between each other. A path between two vertices is a list of consecutive edges connecting the two vertices. A pair of vertices is said to be connected if they have a path between. A graph is said to be connected if each pair of vertices in V has a path between them.

Connectivity of a graph is used to perform a statistical study on whether the vertices in the graph have a path between them. Graph and connectivity of graph have been widely used in a variety of problems for structure modeling, such as brain's functional architecture [65], percolation theory in physics [66], metapopulation [67], and wireless networks [68, 69]. Another typical measure is finding whether the graph is fully connected, which are utilized in these research studies.

One is finding the largest connected cluster in the graph, where cluster means a subgraph that each pair of vertices in the subgraph has at least a link between them (as shown in Fig. 5.1). The growth of the largest cluster is a sufficient trigger for an explosive percolation [67].

The other is finding whether the graph is fully connected, which is defined as follows: Each pair of vertices in the graph has at least a link between them. Moreover, based on the density of links between the vertices, a graph is said to be k-connected ($k = 1,2,3,...$) if for each pair of vertices, there exist at least k mutually independent links between them, a 1-connected graph and a 2-connected graph are shown respectively in Fig. 5.2(a) and (b). Equivalently, a graph is k-connected if and only if the failure of any set of $(k-1)$ vertices will lead to a new isolated vertex in the graph.

5.1.2 Connectivity of Wireless Networks

Wireless networks are usually modeled by graphs, and connectivity is the primary purpose of a wireless network; it links nodes that want to exchange information between them. Most research studies in wireless networks focus on fully connected issues, which ensure that every participant in the wireless networks can be reached by others.

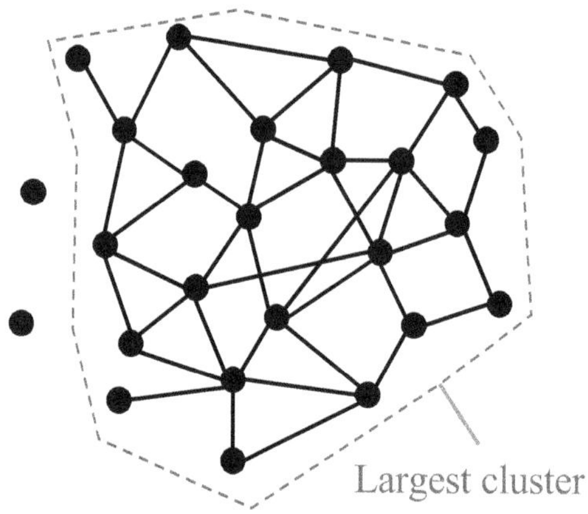

Figure 5.1 The largest cluster of a graph

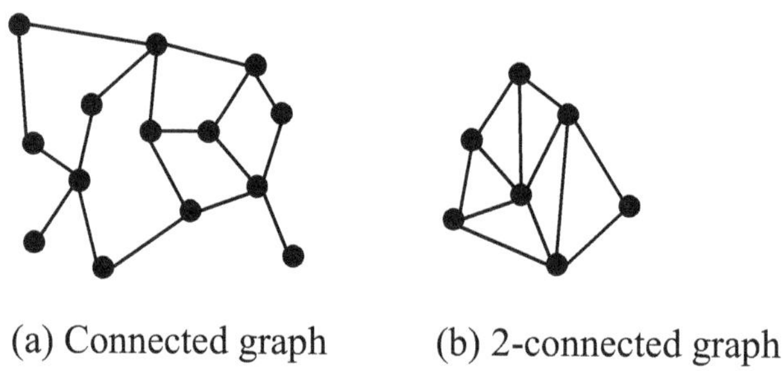

Figure 5.2 An illustration of MI communication link

Connectivity has been widely studied in wireless networks research. Gupta and Kumar [69] derive an expression of range $r(n) = \sqrt{\frac{\log n + c(n)}{\pi n}}$ and prove that $\lim_{n\to\infty} c(n) \to \infty$ is a necessary and sufficient condition for network connectivity (here $c(n)$ is an arbitrary function of n). On the basis of the measure of the average node degree, Bettstetter et al. [70] analyzed the connectivity probability of ad hoc networks. A third sector-based strategy further concerns an angle θ-coverage problem in connectivity analysis [71]. These three main schemes of connecting strategies are summarized by Wang et al. in [72], who further studied the connectivity and capacity problems in the settings of network clustering and node mobility. For realistic wireless networks where the communication range is affected by shadowing, the *effective coverage space* has been proposed for connectivity analysis when the communication range is not deterministic [73, 74]. To be specific, the *effective coverage space* of a node is used to estimate the number of neighbors in order to derive the probability of a randomly picked isolated node. Wang et al. further considered the border effect in the connectivity analysis of the wireless shadowing network [75]. Moreover, the *effective coverage space* method has been applied to connectivity analysis for beamforming networks [76, 77].

In this chapter, we will go through the connectivity of MI networks from the basic system assumption to an optimization problem that aimed to improve the connectivity of MI networks. The systematic analysis methodology of MI networks connectivity is proposed in the following section.

5.2 Connectivity of MI Networks

In this section, we study the connectivity of a large-scale ad hoc MI network whose nodes are randomly located with randomly deployed MI antennas. The pathloss model we use here considers the effect of MI noise via a signal-to-noise ratio (SNR) threshold instead of magnetic signal strength. In addition, the effects of carrier frequency and eddy current are both considered for the determination of signal coverage. To study the MI coverage and connectivity under such assumptions, we develop a Lambert W-function-based integral method to evaluate the effective coverage space and the expected node degree of an MI node. The probability of having no isolated node in the network is further derived to estimate the required parameters for an almost surely connected network. The passive MI waveguide is not considered in this chapter.

Specifically, based on the pathloss model for MI, we explore a carrier frequency selection problem for a maximum communication coverage that benefits the network connectivity. We show that if the carrier frequency is able to change continuously according to different transmitting angle, the coverage space will reach the theoretical maximum under the proposed optimization. In real engineering implementations however, the MI transceiver is not designed for continuous resonant frequency modulation. For such a case, we propose an iterative algorithm to find the optimized carrier frequency, including initialization improvement. Our analytic connectivity analysis is validated by simulations. The impact of carrier frequency and the convergence of the algorithm are illustrated finally. All the notation used here is listed in Table 5.1.

5.2.1 System Assumptions

MI communication is realized by using a couple of coils of wire. According to EM theory, a modulated sinusoidal current in the transmitting coil will induce around it a time-varying magnetic field in space, which further induces a sinusoidal current in the receiving coil. Consequently, this current can be demodulated to information. A single link from a transmitting coil to a single receiving coil is shown in Fig. 5.3. Note that, while this study consider air-cored coils for connectivity analysis, we would like to point out that applying ferrite core with high permeability helps amplify the receiving magnetic signal [78] and thus would improve the connectivity.

5.2.2 Static Magnetic Field

Before we introduce the path loss of MI communication, we investigate the static magnetic field between the transmitter and the receiver in a uniform medium space. The conclusions in this subsection are further used to determine the mutual inductance

Table 5.1 Notation utilized in this paper

Notation	Description
μ	Magnetic Permeability $(H \cdot m^{-1})$
σ	Electric conductivity $(S \cdot m^{-1})$
δ	Skin depth (m)
r	Transmission distance (m)
a_t	Radius of transmitting antenna coil (m)
a_r	Radius of receiving antenna coil (m)
A_t	Area of transmitting coil (m^2)
A_r	Area of receiving coil (m^2)
$\widetilde{N}_t$	Number of turns of transmitting coil
$\widetilde{N}_r$	Number of turns of receiving coil
R	Receiving DC resistance (Ω)
I	Transmitting current (A)
$\mathbf{B}$	Magnetic field strength (T)
M	Mutual inductance (H)
$L_{\text{MI}}(\cdot)$	MI path loss
$\vec{n}_t$	Normal vector of transmitting coil plane
$\vec{n}_{r_i}$	Normal vector of receiving coil plane $(i = 1, 2, 3)$
θ	Angle between observed transmitting direction and $\vec{n}_t$
θ_{r_i}	Angle between observed receiving direction and $\vec{n}_{r_i}$
f	Carrier frequency (Hz)
ω	$\omega = 2\pi f\ (rad \cdot s^{-1})$
P_t	Transmitting power (W)
P_{ih}	Threshold of receiving power (W)
$\mathbf{S}_h$	Threshold of SNR
q_d	Device layer factor
q_m	Medium layer factor

between transmitting and receiving coils. The influence of medium conductivity is considered in the pathloss analysis afterward.

Provided that there is an instantaneous current I in the transmitting coil, the magnetic field strength B at the receiving coil can be considered as a magnetic-dipole field [79], and we have

$$B = \frac{\mu \cdot I \cdot A_t \cdot \widetilde{N}_t}{4\pi \cdot r^3} \sqrt{3\cos^2\theta + 1}, \tag{5.1}$$

where μ represents the permeability of the space (see Fig. 5.3).

If the angle between the axis direction of a receiving coil and the direction of the magnetic field is θ_r, the magnetic field strength successfully received by the receiver will be $B|\cos\theta_r|$. Hence, a single receiving coil cannot guarantee the receiving signal strength even for a strong magnetic field around if the value of $|\cos\theta_r|$ is very small. Therefore, we use a tri-directional receiving antenna consisting of three coils perpendicular to each other as shown in Fig. 5.4. For the MI transmitter, a transmitting coil with a large area and few turns is preferred for a strong transmitting magnetic momentum and low transmitting power. Moreover, we apply a single transmitting coil for

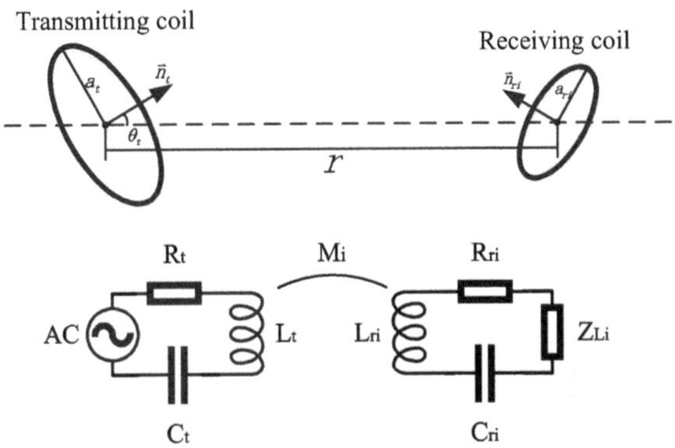

Figure 5.3 An illustration of the MI communication link

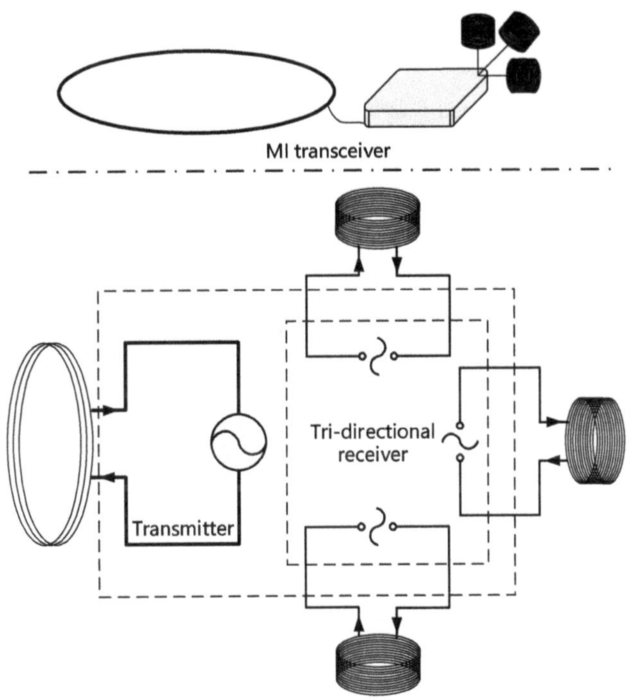

Figure 5.4 An illustration of an MI transceiver

the limitation of space in complex environments. As a result, our considered MI communication system consists of a single-coil transmitting antenna and a tri-directional coils receiving antenna as illustrated in Figs. 5.4 and 5.5. According to our prototype implementation, such a system can realize one-hop communications up to hundreds of meters (with tens of watts of transmission power and tens of meters of transmission antenna) in the very harsh environments, such as coal mines.

For a tri-directional receiving antenna in Fig. 5.5, let $\theta_{r_i}, i = 1, 2, 3$ be the angles between the axis directions of receiving coils and the magnetic field. We further present the following theorem based on [80].

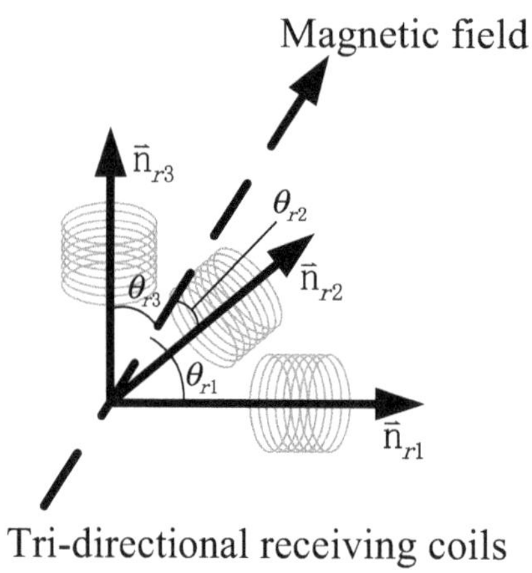

Figure 5.5 Signal receiving with three-directional antenna

Theorem 5.1 *If the magnetic field at a tri-directional MI receiving antenna has a magnitude of B and a random direction $\vec{n}_B$, we have*

$$\begin{aligned}\inf\left(\max\{B|\cos\theta_{r_1}|, B|\cos\theta_{r_2}|, B|\cos\theta_{r_3}|\}\right) \\ = \frac{\sqrt{3}}{3}B.\end{aligned} \tag{5.2}$$

Proof We assume $|\cos\theta_{r_1}| \geq |\cos\theta_{r_2}|$ and $|\cos\theta_{r_1}| \geq |\cos\theta_{r_3}|$, i.e.,

$$\max\{B|\cos\theta_{r_1}|, B|\cos\theta_{r_2}|, B|\cos\theta_{r_3}|\} = B|\cos\theta_{r_1}|. \tag{5.3}$$

Substituting the left side of equation (5.2) with $B\cdot\inf(|\cos\theta_{r_1}|)$, we have the following equation:

$$B \cdot \inf(|\cos\theta_{r_1}|) = \frac{\sqrt{3}}{3}B. \tag{5.4}$$

Because $\vec{n_{r_i}}$ are perpendicular to each other, we have

$$\sum_{i=1}^{3} |\cos\theta_{r_i}|^2 = |\vec{n_B}|^2 = 1, \tag{5.5}$$

which leads to

$$\begin{aligned}\inf\left(|\cos\theta_{r_1}|^2\right) &= \inf\left(1 - |\cos\theta_{r_2}|^2 - |\cos\theta_{r_3}|^2\right) \\ &= 1 - \sup\left(|\cos\theta_{r_2}|^2\right) - \sup\left(|\cos\theta_{r_3}|^2\right) \\ &= 1 - 2|\cos\theta_{r_1}|^2.\end{aligned} \tag{5.6}$$

Equation (5.6) is equivalent to

$$\inf\left(|\cos\theta_{r_1}|^2\right) + 2|\cos\theta_{r_1}|^2 = 1, \tag{5.7}$$

which implies

$$|\cos\theta_{r_1}| \geq \frac{\sqrt{3}}{3}. \tag{5.8}$$

Thus, we have Eq. (5.4).

The similar analysis applies also to $|\cos\theta_{r_2}|$ or $|\cos\theta_{r_3}|$, which completes the proof of Theorem 5.1. □

5.2.3 Path Loss

From [15], the path loss of an MI transmitter and an MI receiver can be approximated by

$$L_{\mathrm{MI}_i} \approx \left|\frac{R}{j\omega M_i}\right|^2, \tag{5.9}$$

where $R = \rho_w \cdot 2\pi a_r$ represents the direct-current (DC) resistance of the receiving coil, $\rho_w(\Omega \cdot \mathrm{m}^{-1})$ is the unit-length resistance of the coil wire, and M_i represents the mutual inductance:

$$\begin{aligned} M_i &= \frac{\widetilde{N}_r \Phi_r}{I} \\ &= \frac{\widetilde{N}_r A_r}{I} B|\cos\theta_{r_i}| \cdot g. \end{aligned} \tag{5.10}$$

Hence, the receiving power of one coil is then denoted by

$$P_{r,i} = P_t \frac{\omega^2}{R^2} M_i^2. \tag{5.11}$$

As mentioned in [81], $g = \exp(-\frac{r}{\delta})$ is an additional loss factor due to eddy currents, where $\delta \approx \sqrt{\frac{2}{\omega\mu\sigma}}$ represents the skin depth for conductive materials. Φ_i is the magnetic flux through the receiving coil. Thus, the lower bound of the receiving signal power by a tri-directional receiving antenna for a given transmitting power P_t is obtained by Theorem 5.1, given as

$$\inf\left(\max\{P_{r,i}\}\right) = \frac{1}{3}\frac{P_t \omega^2 \widetilde{N}_r^2 A_r^2}{R^2 I^2} \cdot g^2 \cdot B^2. \tag{5.12}$$

5.2.4 MI Coverage Space

In this chapter, we assume that the carrier frequency is resonant: $\omega_0 = \frac{1}{\sqrt{L_t C_t}} = \frac{1}{\sqrt{L_{r_i} C_{r_i}}}$, where $L_t \simeq 0.5\mu\pi\widetilde{N}_t^2 a_t$ and $L_{r_i} \simeq 0.5\mu\pi\widetilde{N}_r^2 a_r$ are the self-inductance of transmitting and receiving coils, the capacitance C_t and C_{r_i} are selected according to ω_0, L_t, and L_{r_i}, respectively. According to the thermal noise model for an MI link [82],

$$\mathbf{E}\{P_N(\omega)\} \simeq \frac{1}{2}\frac{4kTR^2}{|j\omega L_{r_i} + \frac{1}{j\omega C_{r_i}} + 2R|^2}, \tag{5.13}$$

where $\mathbf{E}\{P_N(\omega)\}$ represents the noise spectral density, $k \approx 1.38 \times 10^{-23}$ J/K is the Boltzmann constant, and $T = 290$ K is the temperature in Kelvin. We assume the bandwidth to be Δf, the power of noise is further calculated by

$$P_{\text{noise}} = \int_{\omega_0-2\pi\cdot 0.5\Delta f}^{\omega_0+2\pi\cdot 0.5\Delta f} \mathbf{E}\{P_N(\omega)\}\frac{1}{2\pi}d\omega. \tag{5.14}$$

The three coils of the tri-direction receiver are assumed active, and all of their receiving signals are processed until there is a qualified one based on code correction. Let $\mathbf{S}_h = \frac{P_{ih}}{P_{\text{noise}}}$ be a threshold receiving SNR, where P_{ih} is the corresponding minimal receiving power. Under this assumption, a tri-directional receiver can successfully receive the signal if the signal power is large enough for any one of the three directions, i.e., $\inf(\max\{P_{r,i}\}) \geq P_{ih}$. Therefore, combining Eq. (5.12) with (5.1), the boundary of the coverage space of an MI node is expressed as

$$r_{\max}(\theta,\omega) = \frac{3\sqrt{2}\mathbf{W}\left(G_\theta^{1/3}(q_d\cdot q_m)^{1/6}\cdot\omega^{5/6}\right)}{\sqrt{\mu\sigma\omega}}. \tag{5.15}$$

We introduce three symbols q_d, q_m, and G_θ to make this equation more concise, where $q_d = \frac{\widetilde{N}_r^2\widetilde{N}_t^2P_tA_r^2A_t^2}{6^7\pi^2(P_{ih}R^2)}$ and $q_m = \mu^5\sigma^3$ conclude the factors from the device layer and the space medium, respectively. When a certain type of device and transmitting power are decided, q_d is a constant value. The space medium in this chapter is assumed uniform, which implies that q_m is also constant. $G_\theta = \sqrt{3\cos^2\theta+1}$ is a function of θ. Moreover, $\mathbf{W}(\cdot)$ represents the upper branch of the Lambert W-function that is the inverse relation of the function $f(x) = xe^x$, with $x \geq 0$.

Boundary of Coverage Space

Considering the condition of successful receiving: $\inf(\max\{P_{r,i}\}) \geq P_{ih}$ and Eq. (5.12), we have

$$P_{ih} \leq \frac{1}{3}\frac{P_t\omega^2\widetilde{N}_r^2A_r^2}{R^2I^2}\cdot g^2\cdot B^2. \tag{5.16}$$

Recalling that $\delta = \sqrt{\frac{2}{\omega\mu\sigma}}$, we further transform the inequality as follows:

$$\begin{aligned} r^6\cdot e^{\frac{2r}{\delta}} &\leq \frac{(\widetilde{N}_r\widetilde{N}_tA_rA_t\mu\omega G_\theta)^2P_t}{48\pi^2R^2P_{ih}}, \\ \frac{r}{3\delta}\cdot e^{\frac{r}{3\delta}} &\leq \left(\frac{\widetilde{N}_t^2\widetilde{N}_r^2A_t^2A_r^2P_t}{6^7\cdot\pi^2P_{ih}R^2}\cdot\mu^5\sigma^3\right)^{1/6}\omega^{5/6}G_\theta^{1/3}. \end{aligned} \tag{5.17}$$

Since the left-hand side of Eq. (5.17) is a monotone increasing with distance r, the maximum of r can be determined by an inverse function of xe^x, i.e., the Lambert W-function:

$$\frac{r_{\max}}{3\delta} = \mathbf{W}\left(\left(\frac{\widetilde{N}_t^2\widetilde{N}_r^2A_t^2A_r^2P_t}{6^7\cdot\pi^2P_{ih}R^2}\cdot\mu^5\sigma^3\right)^{1/6}\omega^{5/6}G_\theta^{1/3}\right). \tag{5.18}$$

From (5.18), Eq. (5.15) is derived straightforwardly.

5.2.5 Network Model

Spatial Distribution

We consider a network consisting of N MI nodes that are of the same type as shown in Fig. 5.4. The spatial distribution of the nodes follows a homogeneous Poisson point process (P.P.P.) on a 3-D space with a density of ρ nodes per unit space. This process is defined by the following two properties.

(a) The number of nodes U_d in each finite subspace $\mathbf{D}$ with a size of $||\mathbf{D}|| = D$ follows a Poisson distribution, i.e.,

$$P(n \text{ nodes in } \mathbf{D}) = P(U_d = n) = \frac{\lambda^n}{n} e^{-\lambda};\ n \in \mathbf{N}_0, \tag{5.19}$$

where $\lambda = \rho D$ represents the expectation $\mathbf{E}\{U_d\}$ and $\mathbf{N}_0$ is the set for positive integers.

(b) The numbers of nodes N_j in disjoint spaces $\mathbf{D}_j, j \in \mathbf{N}_0$, are the independent random variables, i.e.,

$$\begin{aligned} P\left(N_1 = n_1 \bigcap N_2 = n_2 \bigcap \ldots \bigcap N_k = n_k\right) \\ = \prod_{j=1}^{k} P(N_j = n_j). \end{aligned} \tag{5.20}$$

MI Nodes' Deployment

When MI nodes are deployed, we assume that the direction of their transmitting antennas is uniformly random, i.e.,

$$\begin{aligned} \vec{n}_t &= (\cos\alpha_t \cos\beta_t,\ \cos\alpha_t \sin\beta_t,\ \sin\alpha_t) \\ \alpha_t &\sim \mathcal{U}(0,\pi), \beta_t \sim \mathcal{U}(0,2\pi), \end{aligned} \tag{5.21}$$

where $\mathcal{U}(x,y)$ represents a continuous uniform distribution from x to y.

Since the transmission range of an MI node with a given transmitting power is different with θ, there are three situations for two nodes when they try to connect with each other as illustrated in Fig. 5.6, with transmitting coils shown in the side view.

We observe that nodes in situation (a) are connected, while nodes in situation (c) are not connected. In situation (b) however, two MI nodes can only establish a one-way but not a two-way link. Although we can obtain a directed connection model on the basis of this situation physically, most modern communication protocols need feedback information even for one-way transmission; hence, we treat situation (b) as an unconnected case too.

Assume that two nodes N_1 and N_2 exchange information at a given power P_t and a carrier frequency ω. Vectors $\vec{n}_{1_t}$ and $\vec{n}_{2_t}$ denote the transmitting antenna's directions of N_1 and N_2, respectively, and $\vec{r}$ is the vector starting from point N_1 to N_2. θ_{12} denotes the angle between n_{1t} and $\vec{r}$, while θ_{21} denotes the angle between $\vec{n}_{2_t}$ and $-\vec{r}$, i.e., $\cos\theta_{12} = \frac{\vec{n_{1t}} \cdot \vec{r}}{|\vec{n_{1t}}| \cdot |\vec{r}|}$ and $\cos\theta_{21} = \frac{\vec{n_{2t}} \cdot (-\vec{r})}{|\vec{n_{2t}}| \cdot |\vec{r}|}$. With Eq. (5.15), we use the following definition to describe the connection between two nodes.

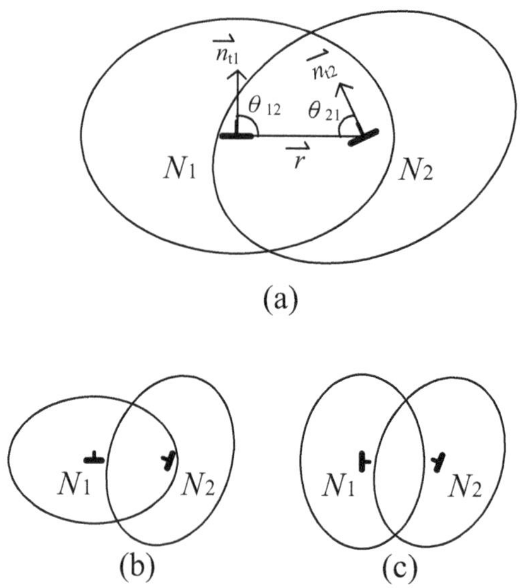

Figure 5.6 Connection model: (**a**) connected, (**b**) unconnected, and (**c**) unconnected [80]

Definition 5.2 For two randomly picked MI nodes N_1 and N_2, they are connected if and only if

$$|\vec{r}| \leq r_{\max}(\theta_{12},\omega) \ and \ |\vec{r}| \leq r_{\max}(\theta_{21},\omega), \tag{5.22}$$

where $r_{\max}(\theta_{12},\omega)$ and $r_{\max}(\theta_{21},\omega)$ are defined according to Eq. (5.15).

5.2.6 Effective Coverage Space

For an MI node, the range is defined by (5.15), and it depends on the antenna angle. The effective two-way transmitting distance of two arbitrary MI nodes N_1 and N_2 is given by

$$\widehat{r} = \min\{r_{\max}(\theta_{12}), r_{\max}(\theta_{21})\}, \theta_{12}, \theta_{21} \in [0, \frac{\pi}{2}], \tag{5.23}$$

where $r_{\max}(\theta_{12})$ and $r_{\max}(\theta_{21})$ are the ranges of N_1 and N_2 and are given by Eq. (5.15) with the same arbitrary ω. The effective coverage space is then given by $\mathbf{E}[D] = \mathbf{E}[\frac{4}{3}\pi\widehat{r}^3]$.

The probability density function (PDF) of $\widehat{r}$ is written as

$$\begin{aligned} &\mathbf{P}(\min\{r_{\max}(\theta_{12}), r_{\max}(\theta_{21})\} = r_{\max}(\theta)) \\ =&\mathbf{P}(\max\{\theta_{12}, \theta_{21}\} = \theta). \end{aligned} \tag{5.24}$$

To solve (5.24), we use the CDF and PDF of θ_{12} and θ_{21} as follows:

(1) The CDF of θ_{12} and θ_{21} equals the ratio of the area of the spherical cap S_θ above the intersection ring over the area of the half sphere as shown in Fig. 5.7 and is given by

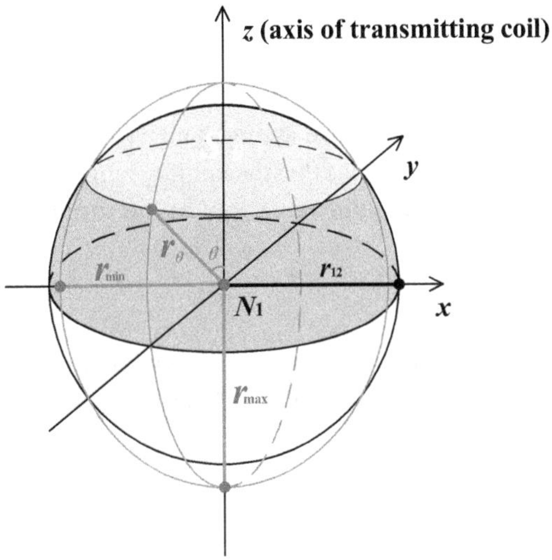

Figure 5.7 Effective transmitting range [80]

$$
\begin{aligned}
F_\theta(\theta_{12}) = F_\theta(\theta_{12}) &= \mathbf{P}(\theta_{12} < \theta) \\
&= \frac{\int_0^\theta 2\pi r_\theta^2 \sin\theta \cdot d\theta}{2\pi r_\theta^2} \\
&= 1 - \cos(\theta).
\end{aligned}
\tag{5.25}
$$

(2) The PDF of θ_{12} and θ_{21} is then derived as

$$
\begin{aligned}
f_\theta(\theta_{12}) &= f_\theta(\theta_{21}) \\
&= \frac{\mathrm{d}F_\theta(\theta_{12})}{\mathrm{d}\theta} = \sin\theta.
\end{aligned}
\tag{5.26}
$$

Using Eqs. (5.25) and (5.26), together with the rule of obtaining the PDF of the maximum of two independent and identically distributed (i.i.d.) random variables, the PDF given in (5.24) is further expressed as

$$
\begin{aligned}
\mathbf{P}(\widehat{r} = r_{\max}(\theta)) &= \mathbf{P}(\max\{\theta_{12}, \theta_{21}\} = \theta) \\
&= f_\theta(\theta_{12})F_\theta(\theta_{21}) + f_\theta(\theta_{21})F_\theta(\theta_{12}) \\
&= 2(1 - \cos\theta)\sin\theta.
\end{aligned}
\tag{5.27}
$$

From (5.27), we derive the expectation $\mathbf{E}[\widehat{r}^3]$ given as

$$
\begin{aligned}
\mathbf{E}(\widehat{r}^3) &= \mathbf{E}(\min\{r_{\max}^3(\theta_{12}), r_{\max}^3(\theta_{21})\}) \\
&= \int_0^{\frac{\pi}{2}} r_{\max}^3(\theta) \cdot \mathbf{P}\Big(\min\{r_{\max}^3(\theta_{12}), r_{\max}^3(\theta_{21})\} = r_{\max}^3(\theta)\Big) \\
&= \int_0^{\frac{\pi}{2}} r_{\max}^3(\theta) \cdot 2\sin\theta(1 - \cos(\theta)) \cdot d\theta.
\end{aligned}
\tag{5.28}
$$

Equation (5.28) provides an integral approach to estimate the *effective coverage space* of the MI node. The above analysis can also be applied to other channel model with different $r_{\max}(\theta)$ functions under a homogeneous P.P.P.

Note that in the above discussions, we have assumed a given carrier frequency. In fact, the value of carrier frequency also affects the transmitting range. The optimization of the carrier frequency w.r.t. the transmitting range will be discussed in Section 5.3.

5.2.7 Connectivity Analysis

For a randomly selected node in a homogeneous P.P.P. network, the probability that it has no connection to any other node is defined by $\mathbf{P}(\text{iso})$ with an expression given by [83]

$$\mathbf{P}(\text{iso}) = \exp\{-\mathbf{E}(U)\}, \tag{5.29}$$

where U denotes the node degree.

In a homogeneous P.P.P. with density ρ in a 3-D space, the expectation of the node degree is

$$\begin{aligned} \mathbf{E}(U) &= \rho\mathbf{E}(D) \\ &= \rho\mathbf{E}(\frac{4}{3}\pi\widehat{r}^3). \end{aligned} \tag{5.30}$$

Thus, we obtain the probability of node isolation by substituting (5.28) into (5.30). Let $\mathbf{P}(\text{non-iso})$ be the probability that none of the nodes in the network is isolated. We have now

$$\mathbf{P}(\text{non-iso}) = (1 - \mathbf{P}(\text{iso}))^N. \tag{5.31}$$

For a homogeneous P.P.P. in a space V without the border effect, we apply the Poisson approximation for (5.31) and have

$$\begin{aligned} \mathbf{P}(\text{non-iso}) &\simeq \exp\left(-\rho V\left(1 - e^{-\mathbf{P}(\text{iso})}\right)\right) \\ &\simeq \exp(-\rho V \cdot \mathbf{P}(\text{iso})). \end{aligned} \tag{5.32}$$

Let $\mathbf{P}(\text{con})$ denote the probability that every node pair in the network has at least one path connecting them. It has been proved in [83] that the upper bound for the network connectivity probability is given by

$$\begin{aligned} &\mathbf{P}(\text{con}) = \mathbf{P}(\text{non-iso}) - \varepsilon, \\ &\text{where } \varepsilon \geq 0 \text{ and } \varepsilon \to 0 \; as \; \mathbf{P}(\text{non-iso}) \to 1. \end{aligned} \tag{5.33}$$

At this point, we have derived an analytic expression for the effective coverage space of MI and obtained the probability of node isolation and network connectivity. In Section 5.3, we will further investigate how to optimize the carrier frequency of MI signals to achieve a longer range and better network connectivity.

Evaluation

To be specific, we investigate the effect of node density, transmitting power, and carrier frequency on network connectivity by intensive simulations. In the simulation, we deploy n MI nodes in a cube space $V = d^3$ that is underground, according to a homogeneous P.P.P., with a density of $\rho = \frac{n}{d^3}\ m^{-3}$. The number of turns of the transmitting coil is $\widetilde{N}_t = 24$, and the coil radius is $a_t = 1\ m$. For the receiving antenna coil, we set $\widetilde{N}_r = 100$ and $a_r = 0.1\ m$. A 17 AWG wire is used for the coils providing a unit length resistance of $\rho_w = 0.0166\Omega \cdot \mathrm{m}^{-1}$. We assume that the underground space is composed of dry soil whose conductivity is $\sigma = 0.01\ S{\cdot}\mathrm{m}^{-1}$, while the permeability of soil is close to that of air: $\mu = \mu_0 = 4\pi \times 10^{-7}\ \mathrm{H}{\cdot}\ \mathrm{m}^{-1}$.

Referring to the field test in [31], we set the transmitting power as $P_{t_0} = 40$ dBm and the receiving power threshold as $P_{ih} = -108.6$ dBm, while the bandwidth $\Delta f = 1$ Hz. For rescue purpose, we focus on range-oriented MI systems (e.g., those with > 100 m range) instead of capacity-oriented ones; hence, we use here a relatively high transmitting power and narrow bandwidth. Note that MI devices in non-rescue applications would work with lower power, more bandwidth, but a much shorter range. For instance, the bandwidth of the MI system proposed in [15] for underground sensor networks is about 2 kHz, with a 10–m single hop range. Equipped with 12 9-V batteries (each with 200 mAh capacity), our MI device is able to transmit 430 20-bit packages carrying critical information like target ID and position. If an MI device transmits a package per 5–10 minutes, its working lifetime is 36–72 hours. We tend to believe that such a system would help a lot in rescue scenarios.

The transmitting range w.r.t. the carrier frequency f and transmitting angle θ is evaluated and plotted in Fig. 5.8, which shows the maximum range is about 144.8 m at frequency $f^* \simeq 4.8$ kHz ($\omega^* \simeq 30.37 \times 10^3$ rad $\cdot$ s^{-1}) and the transmitting angle $\theta = 0$.

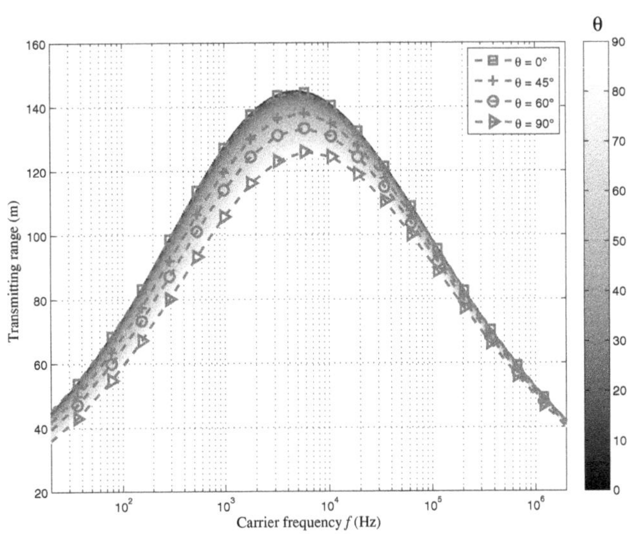

Figure 5.8 Numerical value of the transmitting range. θ ranges from 0° to 90°

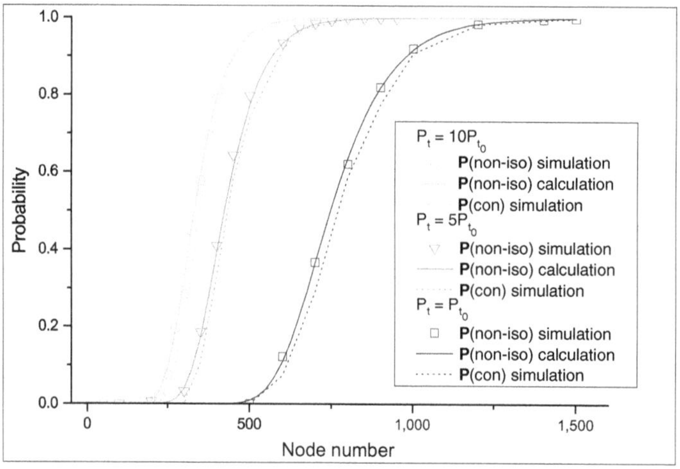

Figure 5.9 Connectivity of MI network A

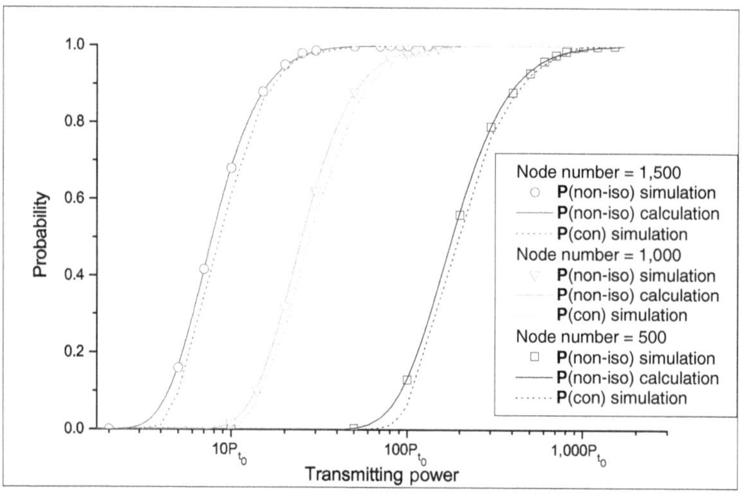

Figure 5.10 Connectivity of MI network B

The antenna direction of the transmitting coil is chosen uniformly among all directions. Two nodes are connected only if they can reach each other either directly or via multi-hop. To avoid the border effect [84], we assume that nodes at the border of the area have MI links via the borderline to nodes on the opposite side of the space.

The simulation is repeated on 1,000 different network deployments in the work of [31] with two additional transmitting power $P_t = 5P_{t0}$ and $P_t = 10P_{t0}$, respectively. For $P_t = P_{t0} = 40$ dBm, more than 1,500 nodes are needed for an almost surely fully connected network in a 1km × 1km × 1km space as shown by the lines with square scatters in Fig. 5.9.

Similarly, the results of **P**(non-iso) and **P**(con) versus the transmitting power level are shown in Fig. 5.10. The side length is changed to $d = 1.5$ km, and we evaluate the connectivity under three different node densities, where a logarithmic coordinate is used for *Pt*.

With these two sets of simulations, we cover a wide range of power and node density. It is observed that the simulation values of **P**(non-iso) (Scatter) in both Figs. 5.9 and 5.10 agree with the theoretical value (Line), which proves the correctness of our connectivity analysis. Note that the value of **P**(con) is slightly smaller than **P**(non-iso) and very close to **P**(non-iso) when they are larger than 0.99, which implies that the network is almost surely connected.

5.3 Carrier Frequency Optimizations

As **P**(non-iso) increases with $\mathbf{E}(D)$, we present the following optimization problem to improve the network connectivity:

$$\begin{aligned} &\arg\max_{\omega} \mathbf{E}(D) \\ &s.t.\ \omega \geq 0. \end{aligned} \tag{5.34}$$

In this subsection, we first provide an ideal *frequency-switching method (FSM)* that maximizes the coverage space $\mathbf{E}(D)$ theoretically. The FSM method may not be feasible for the existing MI devices with determined capacitance. Hence, we further propose a practical *frequency-fixed method (FFM)* that is based on an iterative algorithm.

5.3.1 Frequency-Switching Method

According to Eq. (5.30), the optimization problem (5.34) is equivalent to

$$\begin{aligned} &\arg\max_{\omega} r_{\max}(\theta,\omega) \\ &s.t.\ \omega \geq 0. \end{aligned} \tag{5.35}$$

We start the analysis with the following differentiation of $r_{\max}$ w.r.t. ω, i.e.,

$$\begin{aligned} \frac{\mathrm{d}r_{\max}}{\mathrm{d}\omega} &= \mathbf{W}(G_\theta^{1/3}(q_d q_m)^{1/6}\cdot\omega^{5/6})\cdot \\ &\frac{\left(\mu\sigma\left(2-3\mathbf{W}(G_\theta^{1/3}(q_d q_m)^{1/6}\cdot\omega^{5/6})\right)\right)}{\sqrt{2}(\mu\sigma\omega)^{3/2}\left(1+\mathbf{W}(G_\theta^{1/3}(q_d q_m)^{1/6}\cdot\omega^{5/6})\right)}. \end{aligned} \tag{5.36}$$

Following the single stationary point for $\omega > 0$, we have

$$\omega^* = \left(\frac{\left(\frac{2}{3}\right)^6 e^4}{G_\theta^2(q_d q_m)}\right)^{1/5}. \tag{5.37}$$

One can observe from Eq. (5.36) that $\frac{\mathrm{d}r_{\max}}{\mathrm{d}\omega} > 0$ when $\omega < \omega^*$ and $\frac{\mathrm{d}r_{\max}}{\mathrm{d}\omega} < 0$ when $\omega > \omega^*$, which implies that $r^*_{\max}(\theta) = r_{\max}(\theta,\omega^*)$ is the largest coverage range.

Substituting (5.37) into (5.15), we can express the optimal range without the Lambert W-function:

$$r_{\max}^{*} = \frac{2^{9/10} 3^{3/5} G_{\theta}^{1/5} (q_d q_m)^{1/10}}{e^{2/5} \sqrt{\mu\sigma}}. \tag{5.38}$$

However, this optimized coverage range based on the frequency selection Eq. (5.37) requires the carrier frequency switching continuously between $\omega_{\max}$ and $\omega_{\min}$, which are selected based on Eq. (5.37) for $G_\theta = 1$ and $G_\theta = 2$, respectively, i.e.,

$$\begin{aligned} \omega_{\max} &= \left(\frac{\left(\frac{2}{3}\right)^6 e^4}{q_d q_m} \right)^{1/5}, \\ \omega_{\min} &= \left(\frac{\left(\frac{2}{3}\right)^6 e^4}{4 q_d q_m} \right)^{1/5}. \end{aligned} \tag{5.39}$$

In this process, the capacitance is required to switch continuously, which may not always be feasible for engineering applications. Accordingly, we call FSM an ideal method, which can be regarded as an upper bound for the connectivity analysis.

5.3.2 Frequency-Fixed Method

Now, we consider a more practical method with a fixed frequency for engineering applications. From Eq. (5.28), we see that the effective coverage space $\mathbf{E}(D)$ and its differentiation do not have a closed-form expression. In addition, a numerical calculation of the transmitting range $r_{\max}(\theta, \omega)$ (see Fig. 5.8) further shows that $\mathbf{E}(D)$ is also non-convex. To find the solution to the non-convex problem, we develop an iterative algorithm for optimized frequency searching based on the gradient descent approach as shown in Algorithm 1. The proposed idea is to apply a gradient descent method based on the numerical calculation of $\mathbf{E}(D)$ and $\nabla f(\omega) = -\frac{\mathrm{d}\mathbf{E}(D)}{\mathrm{d}\omega}$. The step length in each iteration is determined by a backtracking line search, where the parameter α is chosen between 0.01 and 0.3, and β is chosen between 0.1 and 0.8 [85].

To improve the speed of the algorithm, we use the following analytical approximation to the Lambert W-function with negligible tolerance [86]:

$$\begin{aligned} W_0^{+} = {} & (1+\epsilon) \ln \left(\frac{6x}{5 \ln[2.4(x/\ln(1+2.4x))]} \right) \\ & - \epsilon \ln \left(\frac{2x}{\ln(1+2x)} \right), \end{aligned} \tag{5.40}$$

where $\epsilon \simeq 0.4586887$.

The initial value of ω has a large impact on the convergence of the algorithm. We further discover that picking the initial value based on FSM enables us to find a good solution with lower complexity: ω among $[\omega_{\min}, \omega_{\max}]$ are optimal for a certain θ and are close to each other as shown in equation (5.37). Therefore, the number of iteration

Algorithm 1: Gradient Descent Algorithm

/* $\widetilde{N}_t$, Number of transmitting coil turns $\widetilde{N}_r$, Number of receiving coil turns P_t, Transmitting power A_t, Area of transmitting coil A_r, Area of receiving coil μ, Permeability σ, Conductivity P_{rh} Threshold of receiving power */

Input: $\widetilde{N}_t, \widetilde{N}_r, P_t, A_t, A_r, \mu, \sigma, P_{rh}$

/* Initialization */

$\omega = \omega_0,\ \omega_0 > 0$

$f(\omega) = -\mathbf{E}(D)$

$\alpha \in [0.01, 0.3]$

$\beta \in [0.1, 0.8]$

while $\nabla||f(\omega)||_2 \geq \eta$ **do**

  $\Delta\omega := -\nabla f(\omega)$.

  /* Backtracking line search. */

  $t := 1 \cdot 10^3 (rad \cdot s^{-1})$

  while $f(\omega + t\Delta\omega) > f(\omega) + \alpha t \nabla f(\omega) \cdot \Delta\omega$ **do**

    $t := \beta t$

  end

  /* Update ω^*. */

  $\omega^* := \omega^* + t\Delta x$

end

can be significantly reduced if we choose ω_0 in $[\omega_{\min}, \omega_{\max}]$. A detailed discussion is provided in Section 5.3.3.

5.3.3 Evaluation

Frequency Selection

With the same settings in 5.2.7, we evaluate the connectivity performance $\mathbf{P}$(non-iso) with ω_{Ideal}, i.e., the ideal FSM based on Eq. (5.37), versus the node numbers from 400 to 1,400 in a 1km × 1km × 1km space under a transmitting power level of $P_t = P_{t_0}$. Here we use this theoretical optimized result as the performance upper bound. We select several representative fixed frequencies and evaluate their $\mathbf{P}(\text{non-iso})\omega$ over $\mathbf{P}(\text{non-iso})\omega_{\text{Ideal}}$ under the same deployment and power conditions. Different selection of parameters α and β will only affect the convergence of algorithm with negligible changes on the solution.

As shown in Fig. 5.11, the y coordinate represents the value of $\frac{\mathbf{P}(\text{non-iso})\omega}{\mathbf{P}(\text{non-iso})\omega_{\text{Ideal}}}$ under different frequency selections. We have $\omega_{\max} \simeq 40.07 \cdot 10^3$ rad $\cdot$ s^{-1} and $\omega_{\min} \simeq 30.37 \cdot 10^3$ rad $\cdot$ s^{-1} based on the above settings, while $\omega^* \simeq 37.16 \cdot 10^3$ rad $\cdot$ s^{-1} is the optimized frequency by Algorithm 1. A frequency gap of $d\Omega = 10 \cdot 10^3$ rad $\cdot$ s^{-1} helps us understand the performance of other carrier frequencies: (1) For either side of ω^*, frequencies closer to ω^* have better performance; (2) for a frequency pair around ω^*, like $\omega^* \pm d\Omega$, higher frequency has a better performance than the lower one.

Since the connectivity performance under the frequencies ω^*, $\omega_{\max}$, and $\omega_{\min}$ are quite close, we present a local frequency selection evaluation in Fig. 5.12, where a smaller frequency gap $d\omega = 10^3$ rad $\cdot$ s^{-1} is used. The result shows that $\omega_{\max}$ has a

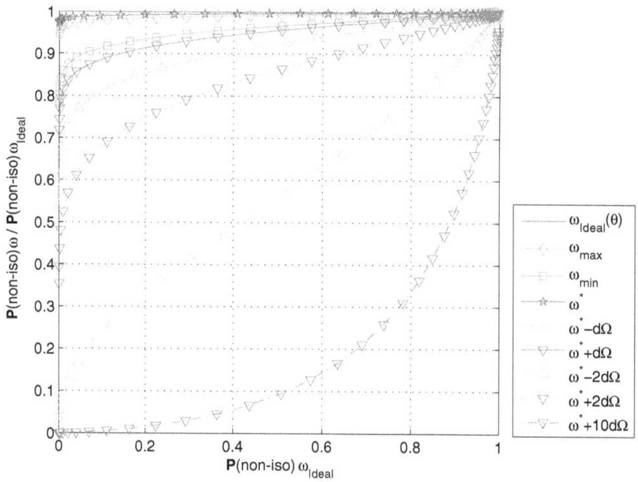

Figure 5.11 **P**(non-iso) under different frequencies

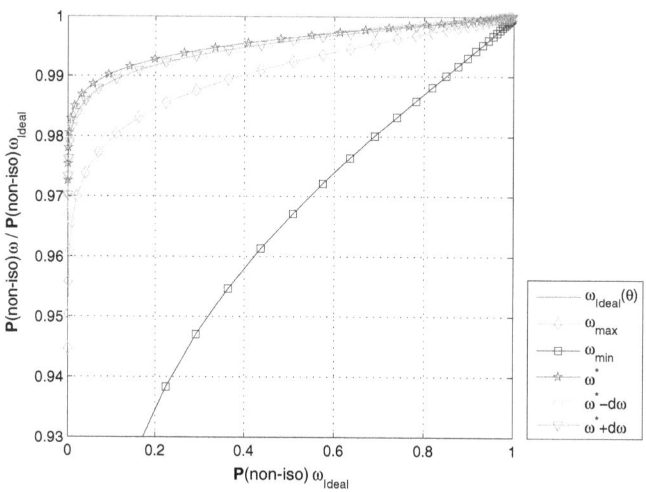

Figure 5.12 **P**(non-iso) under local frequencies

better connectivity than $\omega_{\min}$, which further proves that ω^* is optimal. However, the differences between ω^*, $\omega^* - d\omega$ and $\omega^* + d\omega$ are quite small, which implies that a kHz deviation of frequency selection can be neglected on the perspective of connectivity for practical applications.

Initialization and Convergence Analysis

In the evaluation, we randomly select frequency among $(0, 1]$ $10^6\text{rad} \cdot \text{s}^{-1}$ to initialize Algorithm 1. For backtracking line search, we choose the parameters $\alpha = 0.3$, meaning that we accept a decrease in the target function by 30% and $\beta = 0.8$ for a less crude search, based on which the algorithm will have a fast convergence. The convergence of the algorithm is shown in Fig. 5.13, where the 10 curves are for the 10

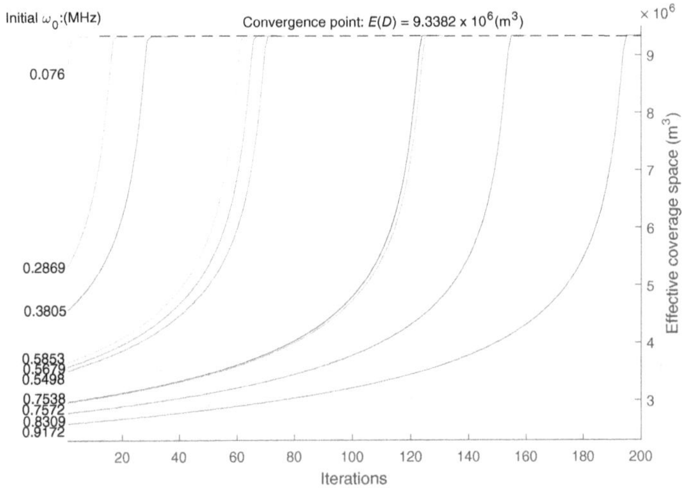

Figure 5.13 Algorithm convergence under random initial ω

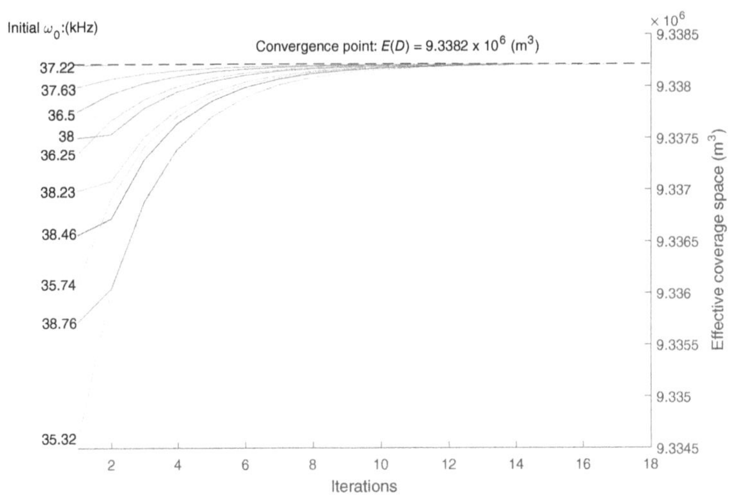

Figure 5.14 Algorithm convergence under random initial $\omega \in [\omega_{\min}, \omega_{\max}]$

randomly selected frequencies for initialization. It is interesting to observe that all the 10 randomly initial selections converge to the same point. In other words, the proposed algorithm is almost sure to achieve the optimal without specific initialization.

As mentioned at the end of Section 5.3.2, we can improve the initialization according to Eq. (5.37) to quicken the convergence of the algorithm. For this purpose, we use 10 random $\omega_0 \in [\omega_{\min}, \omega_{\max}]$ as the initialization frequency, respectively. As shown in Fig. 5.14, the number of iterations is largely reduced, and all 10 initializations converge to the same optimal value within 20 iterations, since the space coverage of these initial frequencies is very close to the optimal. This result proves that the proposed initialization method improves the convergence performance of the algorithm remarkably.

6 MI Network Performance

In this chapter, the network throughput and capacity are derived and analyzed. First, the deployment strategies for MI networks are introduced. Then we present the typical network topologies for MI networks. After that, we compare the performance of different network topologies regarding congestion, node failure, and power consumption, among others.

6.1 Deployment

The deployment of network nodes within a given area of interest is an important aspect of the network design. Since the mutual inductance between two transceiver coils is extremely position- and orientation-selective, the deployment of antenna and network nodes has a significant effect on the MI network performance, especially for the node with a single antenna.

6.1.1 Mutual Inductance in a Common Coordinate System

To study the deployment strategies of the network system, such as the cooperative MI system, the mutual inductance model for every link in a common coordinate system should be derived. The authors of [24] and [25] noticed this problem by deriving the expression of mutual inductance between two arbitrary deployed antennas. As shown in Fig. 6.1, there are a transmitter R and a receiver D in a common Cartesian coordinate xoy. Suppose that the antenna orientation is defined as the angle between the normal vector of the coil and the y axis. The transmitter R is with an arbitrary position (x,y) and an arbitrary antenna orientation θ. The receiver D is with an arbitrary position (x_d, y_d) and an arbitrary antenna orientation θ_d. At the same time, we construct another Cartesian coordinate frame $x_{(r)}Rx_{(r)}$. In this frame, the position of R is the coordinate origin and the normal vector of coil R is the ordinate. Let $\mathbf{v} = (x,y,\theta)$ and $\mathbf{v_d} = (x_d, y_d, \theta_d)$, which denote antenna deployment. According to the magnetostatic field theorems, the field at an arbitrary point is approximately expressed under spherical coordinates with magnetic field components

$$\mathbf{H}_{(r)}\left(r,\theta_{(r)}\right) = \left(\frac{na^2 I_0 \cos\theta_{(r)}}{2r^3}\hat{e}_r + \frac{na^2 I_0 \sin\theta_{(r)}}{4r^3}\hat{e}_\theta\right) \cdot G_e, \tag{6.1}$$

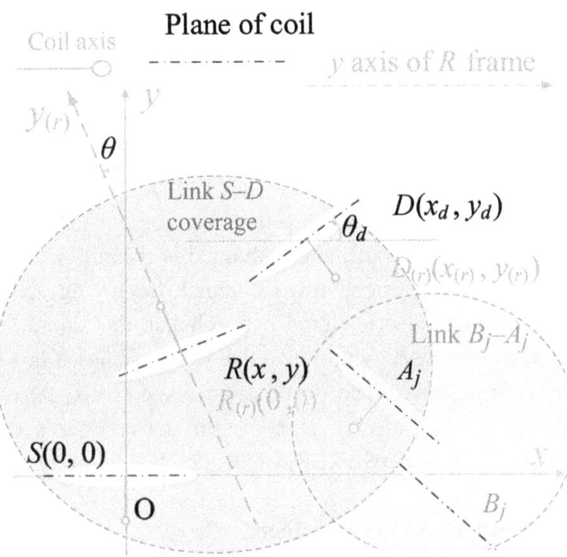

Figure 6.1 A cooperative MI network with a single relay. The subscript((r)) denotes frame R [24]

where $G_e = \exp(-\frac{r}{\delta})$ is an additional loss factor due to the eddy current in which $\delta \approx \sqrt{\frac{1}{\pi f \mu \sigma}}$ according to [32]. The expression of the magnetic field at D under the frame R is deduced as

$$\mathbf{H}_{(r)}\left(x_{(r)}, y_{(r)}\right) = \frac{nI_0 a^2}{4} G_e \left(\frac{2y_{(r)}^2 - x_{(r)}^2}{x_{(r)}^2 + y_{(r)}^2}, \frac{3x_{(r)}y_{(r)}^2}{x_{(r)}^2 + y_{(r)}^2} \right)^T. \tag{6.2}$$

Rotating and translating from frame R to frame S and obtain

$$\begin{pmatrix} x_{(r)} \\ y_{(r)} \end{pmatrix} = \begin{bmatrix} \cos\theta & \sin\theta \\ -\sin\theta & \cos\theta \end{bmatrix} \begin{pmatrix} x_d - x \\ y_d - y \end{pmatrix}, \tag{6.3a}$$

$$\mathbf{H}(x_d, y_d) = \begin{bmatrix} \cos\theta & \sin\theta \\ -\sin\theta & \cos\theta \end{bmatrix}^{-1} \mathbf{H}_{(r)}. \tag{6.3b}$$

By substituting (6.2) and (6.3a) into (6.3b), we obtain the expression $\mathbf{H}(\mathbf{v}, \mathbf{v}_d)$. Thus, the mutual inductance between the arbitrarily placed coils R and D w.r.t. their antenna deployments $\mathbf{v}$ and $\mathbf{v}_d$ is $\partial\Phi/\partial I_0$, i.e.,

$$\begin{aligned} M_{RD}(\mathbf{v}, \mathbf{v}_d) &= \frac{\partial \left(n\mu\pi I_0 a^2 e^{-\frac{r}{\delta}} \left| \mathbf{H}(\mathbf{v}, \mathbf{v}_d) \cdot \mathbf{n}_d(\mathbf{v}_d) \right| \right)}{\partial I_0} \\ &= \frac{\left(\Delta x^2 + \Delta y^2\right)^2 \cos(\theta_d - \theta) - 6\Delta x \Delta y \sin(\theta_d + \theta) - 3\left(\Delta x^2 - \Delta y^2\right)^2 \cos(\theta_d + \theta)}{2\kappa^{-1} \cdot e^{\sqrt{\Delta x^2 + \Delta y^2}/\delta} \sqrt{\left(\Delta x^2 + \Delta y^2\right)^5}}, \end{aligned} \tag{6.4}$$

where $\Delta x = (x_d - x)$, $\Delta y = (y_d - y)$, $\mathbf{n}_d(\mathbf{v}_d)$ is the normal vector of the receiving coil plane, $\kappa = \mu\pi n^2 a^4/4$, μ denotes the medium permeability, and Φ denotes the magnetic flux.

6.1.2 Deployment of Active Relay

For the EM waves communication networks, the cooperative relay techniques have been widely investigated to increase the communication performance. Researchers use the spatial diversity techniques that often rely time-varying channel conditions that might reduce the transmission SER and outage probability. Unlike the EM channel, the MI channel condition is quasi-static time varying for the lack of multipath effect. Also, the outage probability of the MI channel is meaningless. However, due to the position- and orientation-selective MI channel, deploying active relay with the proper position and antenna orientation can enhance the signal strength. Suppose a typical cooperative network with one active relay as shown in Fig. 6.1 where S, R, and D are the source, relay, and destination, respectively. Also, there exists another active node A_j. Let $\mathcal{N} = \{S,R,D,A_j\}$ is the active node set in the coverage of the S. According to this network, we introduce two relay schemes for relay deployment.

(1) Amplify-and-forward (AF) relay scheme

In AF relaying schemes, the relay amplifies the analog signal received from the source and forwards it to the destination (without explicitly decoding or demodulating the messages or symbols) within two time slots. During the first time slot, the source broadcasts s to the destination D and relay R; see Fig. 6.1. According to [24] as shown in Fig. 6.2, since the distance between two nodes is sufficiently large, the MI waveguide effect of unexpected relays (e.g., the node A_j) can be ignored. Therefore, the channel power gain of the link S–R and that of S–D are approximately expressed by

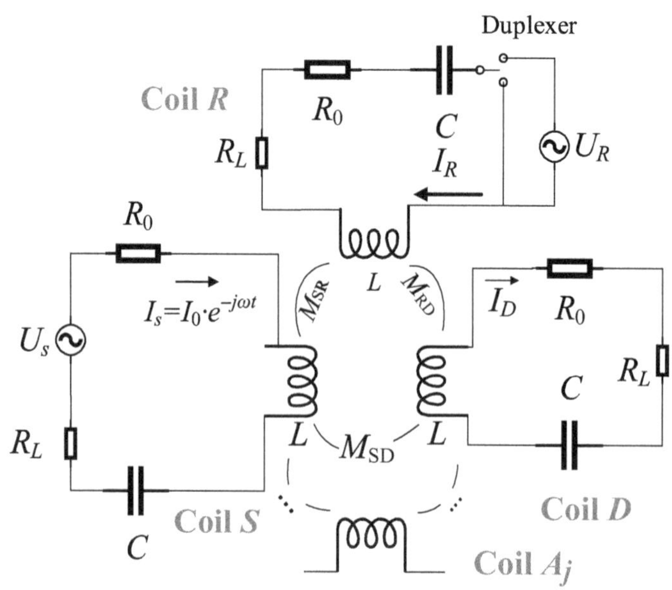

Figure 6.2 Equivalent circuit model for Fig. 6.1 [24]

$$
\begin{aligned}
|H_{SR}(\mathbf{v},f)|^2 &\simeq \frac{(2\pi f)^2 R_L}{|Z_{tr}(f)|^3} \cdot M_{SR}^2(\mathbf{v}) \text{ and} \\
|H_{SD}(f)|^2 &\simeq \frac{(2\pi f)^2 R_L}{|Z_{tr}(f)|^3} \cdot M_{SD}^2,
\end{aligned}
\tag{6.5}
$$

respectively. Thus, the SNR at the destination is

$$
\Upsilon_1(f) = \frac{P_{Sf}(f)}{P_{Nf}} |H_{SD}(f)|^2 = \frac{P_{Sf}(f)}{P_{Nf}} \frac{(2\pi f)^2 R_L}{|Z_{tr}(f)|^3} M_{SD}^2, \tag{6.6}
$$

where $P_{Sf}(f)$ is the transmit power spectral density (PSD), and $Z_{tr} = j2\pi fL + \frac{1}{j2\pi fC} + R_0 + R_L$ is the total impedance of the transceiver circuit.

During the second time slot, the relay R becomes a transmitter. The relay R forwards signals with transmit PSD $P_{Rf}(f)$. The channel power gain of the link R–D is approximately expressed by

$$
|H_{RD}(\mathbf{v},f)|^2 \simeq \frac{(2\pi f)^2 R_L}{|Z_{tr}(f)|^3} \cdot M_{RD}^2(\mathbf{v}). \tag{6.7}
$$

Suppose that the amplification coefficient is β; then the SNR at the destination in the second time slot is given by [87]

$$
\Upsilon_2(f,\mathbf{v}) = \frac{P_{Sf}(f)\beta^2 |H_{SR}|^2 |H_{RR}|^2 |H_{RD}|^2}{\beta^2 |H_{RR}|^2 |H_{RD}|^2 P_{Nf} + P_{Nf}}, \tag{6.8}
$$

where P_{Nf} is the noise PSD and

$$
|H_{RR}|^2 = \left| \frac{Z_{tr}}{Z_{tr}^2 + (2\pi f)^2 \left(M_{SR}^2 + M_{RD}^2\right)} \right| R_L \simeq \frac{R_L}{|Z_{tr}|}.
$$

The amplification coefficient β can be optimized by [24]

$$
\beta(f) = \sqrt{P_{Rf}(f)/(P_{Sf}(f)|H_{SR}|^2 + P_{Nf})|H_{RR}|^2}.
$$

By using maximum ratio combining (MRC), the SNR of the network with a relay is obtained as

$$
\begin{aligned}
\Upsilon_{\mathrm{mrc}}(f,\mathbf{v}) &= \Upsilon_1 + \Upsilon_2(f,\mathbf{v}) \\
&= \frac{P_{Sf}(f)}{P_{Nf}} H_0(f) M_{SD}^2 \\
&\quad + \frac{\frac{P_{Rf}(f)}{P_{Nf}} |H_0(f)|^2 M_{RD}^2(\mathbf{v}) \cdot \frac{P_{Sf}(f)}{P_{Nf}} M_{SR}^2(\mathbf{v})}{\frac{P_{Rf}(f)}{P_{Nf}} |H_0(f)| M_{RD}^2(\mathbf{v}) + \frac{P_{Sf}(f)}{P_{Nf}} |H_0(f)| M_{SR}^2(\mathbf{v}) + 1}, \\
|H_0(f)| &= \frac{(2\pi f)^2 R_L}{|Z_{tr}(f)|^3} = \frac{(2\pi f)^2 R_L}{|(j2\pi fL + \frac{1}{j2\pi fC}) + R_L + R_0|^3}.
\end{aligned}
\tag{6.9}
$$

Through such SNR, we can derive the 3-dB bandwidth by solving the SNR equation

$$
\Upsilon(f) = \frac{1}{2}\Upsilon(f_0), \tag{6.10}
$$

where $f_0 = 1/2\pi\sqrt{LC}$ is the resonance frequency. Suppose that the positive solutions of (6.10) are f_1 and f_2, and the bandwidth is $B_{\rm AF} = |f_2 - f_1|$, i.e.,

$$\begin{aligned} \varpi(\mathcal{Z}_c) &= 2\pi^2C^2f_0^4\mathcal{Z}_c^{-2/3} - 8\pi^2C^2f_0^4R_0^2 + f_0^2, \\ \varrho(\mathcal{Z}_c) &= \sqrt{\left[f_0^2 - 2\pi^2C^2f_0^4\left(4R_0^2 - \mathcal{Z}_c^{-2/3}\right)\right]^2 - f_0^4}, \\ B_{\rm AF}(\mathcal{Z}_c) &= \sqrt{\varpi(\mathcal{Z}_c) + \varrho(\mathcal{Z}_c)} - \sqrt{\varpi(\mathcal{Z}_c) - \varrho(\mathcal{Z}_c)}, \end{aligned} \tag{6.11}$$

where

$$\mathcal{Z}_c = \frac{\begin{array}{c}\{\hbar^2A_0^2\left(M_{RD}^2 + M_{SR}^2\right)^2 + 2\hbar A_0 \cdot \\ \left[M_{SD}^2\left(M_{SR}^2 + M_{RD}^2\right) + 2M_{SR}^2M_{RD}^2\right] \\ + M_{SD}^4\}^{1/2} + \hbar A_0\left(M_{RD}^2 + M_{SR}^2\right) - M_{SD}^2\end{array}}{2\hbar\left[M_{SD}^2\left(M_{SR}^2 + M_{RD}^2\right) + M_{SR}^2M_{RD}^2\right]}, \tag{6.12}$$

$\hbar = (2\pi f_0)^2 R_L \frac{P_{Sf}}{P_{Nf}}$ and $A_0 = \frac{\Upsilon_{\rm mrc}(f_0)}{m\hbar}$. The capacity of the cooperative MI network with an AF relay is given by

$$C_{\rm AF}(\mathbf{v}) = \frac{1}{2}\int_{f_0-\frac{1}{2}B_{\rm AF}(\mathbf{v})}^{f_0+\frac{1}{2}B_{\rm AF}(\mathbf{v})} \log_2(1 + \Upsilon_{\rm mrc}(f, \mathbf{v}))df.$$

Equation (6.12) indicates that the deployment strategy has an effect on the bandwidth and capacity.

(2) Decode-and-forward (DF) relay scheme

The process of the DF relay scheme is also achieved via two time slots. During the first time slot, S broadcasts its signal to D and R, and the relay R decodes the received signal with the data rate $C_{\rm I}$, which can be given by

$$\begin{aligned} C_{\rm I}(\mathbf{v}) = C_{SR}(\mathbf{v}) &= \int_{B_{\rm DF}} \log_2\left(1 + P_{Sf}\frac{H_{SR}(f, \mathbf{v})}{P_{Nf}}\right)df, \\ &= \int_{f_0-\frac{1}{2}B_{\rm DF}}^{f_0+\frac{1}{2}B_{\rm DF}} \log_2\left(1 + P_{Sf}\frac{H_0(f)M_{sr}^2(\mathbf{v})}{P_{Nf}}\right)df, \end{aligned} \tag{6.13}$$

where $B_{\rm DF} = B_{\rm AF}([\frac{1}{2}(R_0 + R_L)]^3)$ derived from $|\Upsilon_1(f)| = \frac{1}{2}|\Upsilon_1(f_0)|$. During the second time slot, R forwards the decoded signal to D. By using MRC, the equivalent data rate in phase II is $C_{\rm II}$, which is

$$C_{\rm I}(\mathbf{v}) = C_{\rm II}(\mathbf{v}) = \int_{f_0-\frac{1}{2}B_{\rm DF}}^{f_0+\frac{1}{2}B_{\rm DF}} \log_2\left(1 + H_0(f)\frac{P_{Rf}M_{RD}^2(\mathbf{v}) + P_SM_{SD}^2}{P_{Nf}}\right)df. \tag{6.14}$$

Different from the AF scheme, the encode–decode process would bring the additional data rate loss due to the time delay, and the total achievable rate of the cooperative network with a DF relay is expressed by [25]

$$Q_{\text{DF}}(\mathbf{v}) = [\iota_{\text{DF}}(\mathbf{v})C_{\text{DF}}(\mathbf{v}) - \delta_{\text{DF}}]^{+}, \tag{6.15a}$$

$$C_{\text{DF}}(\mathbf{v}) = \frac{\Gamma_{\text{fec}}}{2}\min\{C_{\text{I}}(\mathbf{v}), C_{\text{II}}(\mathbf{v})\}, \tag{6.15b}$$

where Γ_{fec} is the gap between the actual data rate and Shannon's capacity, the constant $\delta_{(\cdot)}$ is the data rate loss brought by the time delay, $\iota_{\text{DF}} = 1 - p^{e}_{\text{DF}}$, where p^{e}_{DF} represents the BER. Therefore, the antenna deployment $\mathbf{v}$ also has an effect on the throughput of the cooperative network with a DF relay.

6.1.3 Antenna Deployment Optimization

According to (6.13) and (6.15), the achievable rate of the active relay system can be determined by the relay antenna deployment $\mathbf{v}_r = (x_r, y_r, \theta_r)$. However, many deep-penetrate MI communications are applied to the underground tunnel, which means that the relay can only be deployed in the reachable space. Antenna deployment optimization can remarkably improve the network throughput. Here, the optimization problem of the antenna deployment of DF relay is formulated. Then, an antenna deployment algorithm is proposed to solve this problem.

Suppose the cooperative MI network is as shown in Fig. 6.3 where there are the source S and destination D in the tunnel. The relay R is deployed to obtain the optimal achievable rate gain.

First, to formulate the optimization problem, the achievable rate of the cooperative network with a DF relay is normalized as

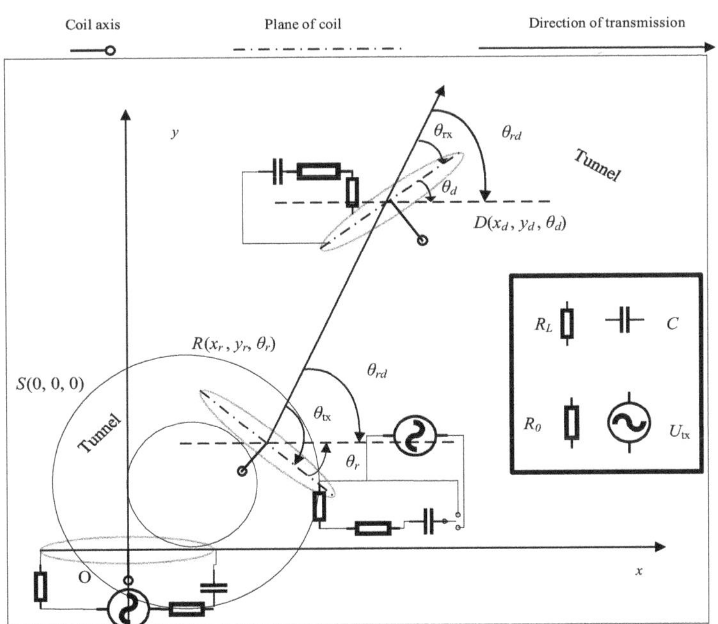

Figure 6.3 Deployment of the DF relay in underground tunnels [25]

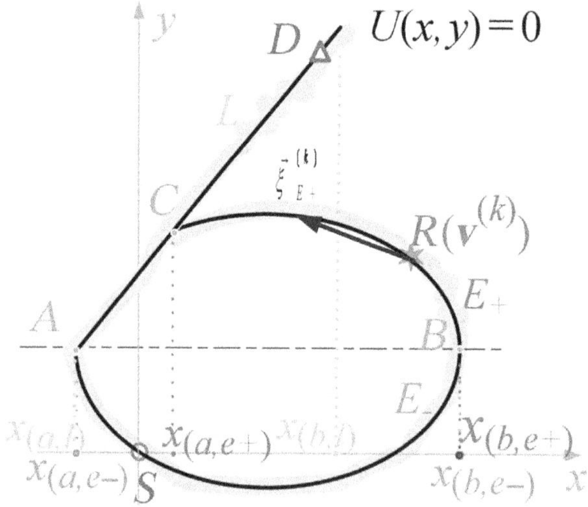

Figure 6.4 Underground tunnel modeling [25]

$$
\begin{aligned}
G(x_r, y_r, \theta_r) &= \frac{\left[\iota_{\mathrm{DF}}^{*} C_{\mathrm{DF}}(x_r, y_r, \theta_r) - \delta_{\mathrm{DF}}\right]^{+}}{Q_{sd}}, \\
Q_{sd} &= \left[\iota_{sd}\Gamma_{\mathrm{fec}} \int_{B} \log_2\left(1 + \tfrac{P_{\mathrm{tot}}}{P_N} H_0(f) M_{sd}^2\right) df - \delta_{sd}\right]^{+},
\end{aligned}
\tag{6.16}
$$

where Q_{sd} is the achievable rate of the MI link S–D. Such a normalized achievable rate is called the DF quasi-gain.

Second, since any tunnel can be fitted into an arc of an ellipse or a straight line segment [25], the tunnel line can be approximately modeled as the combination of n_e arcs of ellipses and n_l straight line segments; see Fig. 6.4 where $n_e = 1$ and $n_l = 1$ without loss of generality. In other words, the tunnel can be expressed by

$$
\mathcal{Y}_i(x_r) = \begin{cases} y_c + R_b \sqrt{1 - \frac{(x_r - x_c)^2}{R_a^2}} & i = E_+, \\ y_c - R_b \sqrt{1 - \frac{(x_r - x_c)^2}{R_a^2}} & i = E_-, \\ \eta x_r + \beta & i = L, \end{cases}
\tag{6.17}
$$

where R_a, R_b, and (x_c, y_c) are the long (short) axis, the short (long) axis, and the center of the ellipse, respectively.

Third, the antenna deployment optimization problem is formulated as

$$
\mathcal{P}_i : \min_{x_r, \theta_r} g_i(x_r, \theta_r) = \min_{x_r, \theta_r} g(x_r, \mathcal{Y}_i(x_r), \theta_r) \tag{6.18a}
$$

$$
s.t.\ x_{(a,i)} \le x_r \le x_{(b,i)}, \tag{6.18b}
$$

$$
\arg \min_{x_r, \theta_r, i} \{\text{sub-problem } \mathcal{P}_i\}. \tag{6.18c}
$$

Algorithm 2: Antenna optimization algorithm.

Input: ε_a is the threshold for the large step vector.
/* $v_r[m][k] = (x_r^{(k)}, \theta_r^{(k)})$ for m-th prober */
while $k{+}{+} < MaximalIterations$ **do**
  for $m = 1$ *to* $\mathcal{M}$ *do* **do**
    $\widehat{g_i}^*(v_r^*) = \min_{m' < \mathcal{M}} g_i(v_r[m'][k])$
    obtain $\vec{\xi}_i^{(k)}$ according to (6.20);
    update $v_r[m][k+1]$ according to (6.19);
  end
end
return the optimal solutions $\widehat{g_i}^*$ and v_r^*

The problem (6.18) can be solved by Algorithm 2 that is based on the random-search and descent methods. In this algorithm, the descent method is based on

$$v_r^{(k+1)} = v_r^{(k)} + \vec{\xi}_i^{(k)} \text{ if } g_i(v_r^{(k+1)}) < g_i(v_r^{(k)}), \tag{6.19}$$

where $\vec{\xi}_i^{(k)} = t_1 \Delta {v_r}^{(k)}$ is the searching vector along the subtunnel i, k represents the kth iteration step, t_1 is the step size, and $v_r^{(k)} = (x_r^{(k)}, \theta_r^{(k)})$. The random-search method is based on

$$\vec{\xi}_i^{(k)} = \begin{cases} \frac{\min\{k,\lambda_a\}}{k}(\chi\hbar_{ba}, 90^\circ\vartheta) & \frac{\widehat{g_i}^*(v_r^*) - g_i(v_r^{(k)})}{\widehat{g_i}^*(v_r^*)} < \varepsilon_a, & (6.20\text{a}) \\ (\chi\hbar_{ba}, 90^\circ\vartheta) \otimes \left(\varpi\Delta_i^{\nabla} v_r^{(k)} + \Delta_i^{\diamond} v_r^{(k)}\right) & \text{otherwise}, & (6.20\text{b}) \end{cases}$$

$$\hbar_{ba} = x_{(b,i)} - x_{(a,i)}, \quad i \in \{E_+, E_-, L\}, \tag{6.20c}$$

where χ and ϑ are the random variables between -0.5 and 0.5, the constant $|\varepsilon_a| \leq 1$, λ_a is the minimal iterations guaranteeing the searching vector with large step size, ϖ is the weight of the subgradient direction, $\otimes$ denotes the element-wise vector multiplication, $\Delta_i^{\nabla} {v_r}^{(k)} = \frac{\nabla g_i(v_r^{(k)})}{||\nabla g_i(v_r^{(k)})||}$ denotes the subgradient direction, and $\Delta_i^{\diamond} v_r^{(k)} = \frac{v_r^* - v_r^{(k)}}{||v_r^* - v_r^{(k)}||}$ denotes the vector pointing to the global optimum.

Next, we discuss the convergence performance of Algorithm 2. Suppose the number of coil turns $n = 10$, the coil radius $a = 0.5$ m, the unit length resistance of the coil $\rho_w = 0.0166\ \Omega/\text{m}$, the load resistor $R_L = R_0 = 2\pi a n \rho_w$. Suppose the resonance frequency $f_0 = 10$ kHz, $\Gamma_p = 0.5$, $\Gamma_{\text{fec}} = 0.5$, $\delta_{\text{DF}} = 2\delta_{sd} = 6.4$, and transmitting PSD, $P_{Rf} = P_{Sf}$. The noise level is -103 dBm, $\mathbf{v}_d = (30\text{ m}, 52\text{ m}, 30^\circ)$ which is 60 m away from S. E_+ and E_- are with parameters ($x_c = 8.91$ m, $y_c = 12.67$ m, $R_a = R_b = 15.49$ m, $x_{a,e+} = x_{a,e-} = -6.58$ m, $x_{b,e+} = x_{b,e-} = 24.4$ m).

Figure 6.5 shows the convergence of Algorithm 2 and GD algorithm under the equality constraint E_+. It can be seen that all $\mathcal{M} = 20$ probers converge to the same point in Fig. 6.5, which means that Algorithm 2 has an excellent global search ability. In addition, most prober lines drastically drop down to the convergence point within 125 iterations.

Finally, we propose an example of the deployment strategy of the antenna of the DF relay. As shown in Fig. 6.6, by using Algorithm 2, we first obtain the optimal

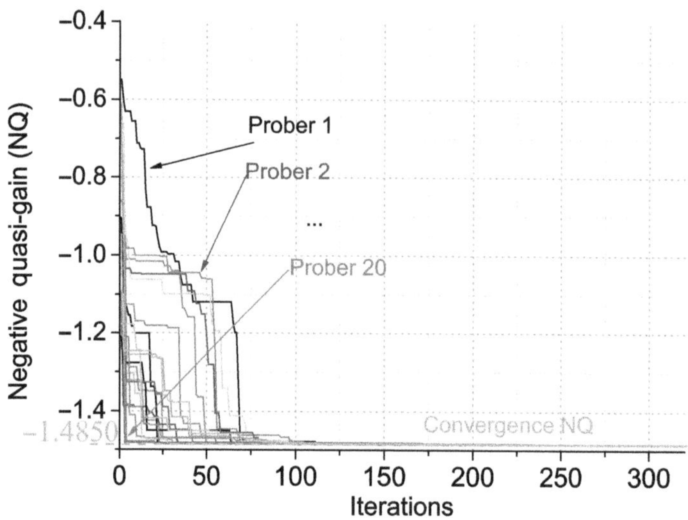

Figure 6.5 Convergence of antenna optimization algorithm [25]

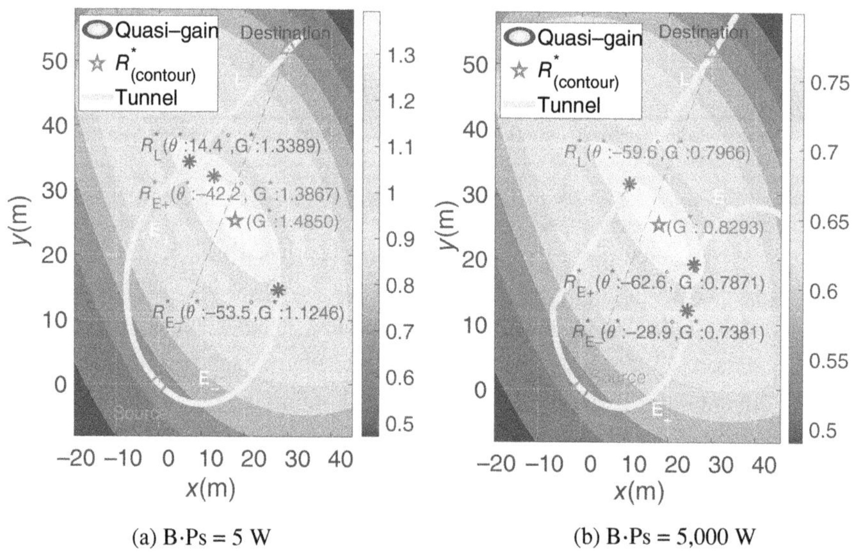

Figure 6.6 Relay deployment by the antenna optimization algorithm. (**a**) Transmit power is 5 W. (**b**) Transmit power is 5 kW [25]

deployments R^*_{E+}, R^*_{E-}, and R^*_L within the subtunnel $E+$, $E-$, and L, respectively. Here, R^*_{E+}, R^*_{E-}, and R^*_L are within the subtunnels $E+$, $E-$, and L, respectively. Then, we deploy the relay to the maximal one of the antenna deployments $\{R^*_{E+}$, R^*_{E-}, and $R^*_L\}$, i.e., R^*_{E+} in Fig. 6.6(a) whose $G^* > 1$. In other words, such deployment of relay can improve the data rate performance. In a few cases (Fig. 6.6(b)) with strong signals, if the quasi-gain of the global optimal antenna deployment (R_L^* in Fig. 6.6(b)) is smaller than 1, we use the direct MI communication as the relay cannot improve data rate performance.

6.2 MI Waveguide Deployment

6.2.1 Introduction

In challenging environments, such as underground, mines/tunnels, and oil reservoirs, the wireless networks enable a large variety of novel and essential applications, including intelligent irrigation, earthquake and landslide forecast, mine disaster prevention and rescue, underground pipeline monitoring, oil recovery, and concealed border patrol, among others [5, 88, 89]. Despite the potential advantages, due to the hostile transmission medium of challenging environments, the well-established existing wireless networks do not work well [88].

Most existing wireless networks use the EM waves for signal propagation. However, the EM waves encounter two major problems in challenging environments, including the extremely small communication range and the highly dynamic channel conditions [88]. To address the above problems, in [90, 91], we developed the *MI waveguide* technique. As shown in Fig. 6.7, wireless communications are accomplished by the consecutive MI between the adjacent MI relay coils. The sinusoidal

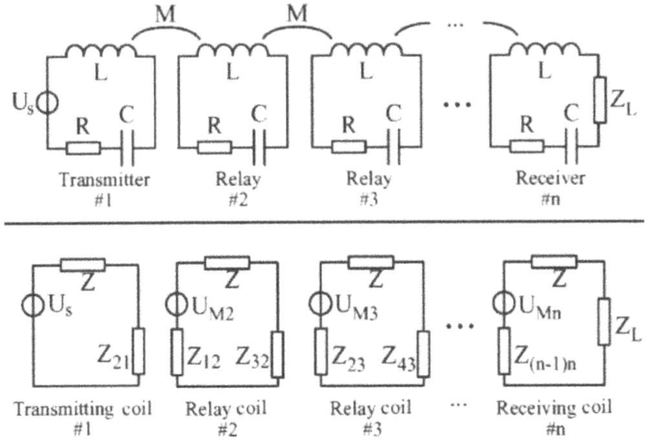

Figure 6.7 The structure and the communication range of an MI waveguide [92]

current in the transmitter coil first creates a time-varying magnetic field, which can induce another sinusoidal current in the first relay coil, so on and so forth. By this way, the MI can be passively relayed by the multiple coils until reaching the MI receiver, forming the so-called MI waveguide. The MI waveguide technique solved the problems of EM wave-based techniques and provided more favorable conditions for wireless networks in challenging environments: (1) The communication range is greatly enlarged. For example, in the soil medium, the range increases from less than 4 m to more than 100 m [90]. (2) The MI channel conditions remain constant in most transmission media, since the attenuation rate of the magnetic fields does not change in nonmagnetic media. (3) The MI relay coils do not consume extra energy, and the unit cost is neglectable. Those coils are easy to deploy and do not need regular maintenances. (4) The system lifetime can be greatly prolonged since the MI-based devices can be recharged wirelessly using the inductive charging technique.

Although the MI waveguide technique solves the point-to-point communication problem in challenging environments, deploying the MI waveguides to connect a large number of wireless nodes is challenging. Specifically, due to the hostile transmission medium, all nodes in the network are isolated unless they are connected by MI waveguides. Hence, the MI waveguide deployment strategy design is essential to realize the MI-based wireless communications among multiple MI nodes. On the one hand, a nontrivial number of relay coils are required to construct a connected and robust wireless network. On the other hand, the intensive deployment of the coils in challenging environments costs a great amount of labor. Therefore, the objective of the deployment strategy in MI-based wireless networks is to *construct a connected and robust network with as few relay coils as possible*. To achieve this goal, three fundamental deployment questions need to be addressed:

- *What is the topology and position of each MI waveguide?* Unlike the disk-shaped communication range of the traditional wireless devices, the range shape of the MI waveguide is more complicated due to the consecutive passive magnetic inductions, as shown in Fig. 6.7. Hence, the deployment strategies in the MI-based networks are completely different. The topology and the position of each MI waveguide need to be carefully designed.
- *How many relay coils are needed?* The waveguide topology/position and the number of relay coils constituting each MI waveguide should be jointly designed to minimize the total relay coil number while guaranteeing the expected network robustness.
- *How robust is the constructed MI-based wireless network?* Since the communication success of the MI waveguide relies on multiple resonant relay coils, the functionality of the MI-based network depends not only on the wireless nodes but also on all the relay coils. The system robustness to the node malfunction and the coil position deviation need to be characterized.

To the best of our knowledge, these questions have not been addressed so far. Moreover, instead of being limited to heuristics, it is more important to develop the deployment solutions that can be rigorously proved to be optimal in terms of robustness and low cost.

In this section, we theoretically investigate the optimal MI waveguide deployment strategies for the MI-based wireless networks in challenging environments. In particular, we start with the MI waveguide deployment in one-dimensional (1-D) networks. The optimal number of relay coils for a 1-D MI link is derived to capture the effects of multiple system parameters, including the transmission distance, operating frequency, coil size, and coil deviations. Then we focus on the deployment strategy for the two-dimensional (2-D) MI networks. Since the 2-D network is constituted by 1-D links, the 2-D deployment strategy is based on the strategy in 1-D networks. We consider the 2-D networks where the wireless nodes are either randomly distributed or located on a regular lattice. We propose an MI waveguide deployment algorithm by utilizing the Voronoi diagram and the Fermat point [93], which is proved to be optimal to trade off between the total number of relay coils and the network robustness to device failure and coil displacement. The effectiveness of the proposed deployment strategy is validated by both theoretical deduction and computer simulations.

The remainder of this section is organized as follows. In Section 6.2.2, the related works are introduced. In Section 6.2.3, the optimal MI waveguide deployment strategy is developed in both 1-D and 2-D networks. Then, in Section 6.2.4, numerical studies are performed.

6.2.2 Related Work

The MI-based communication technique was first introduced as an alternative to the Bluetooth in [94]. The rapid attenuation of the magnetic field strength is exploited to eliminate the mutual interference. However, the high path loss is obviously not an advantage in the proposed applications in challenging environments. The MI waveguide technique is first developed in [95, 96], which is designed as artificial delay filters, dielectric mirrors, distributed Bragg reflectors, slow-wave structures, and coupled cavities, among others. In [90, 91], we first utilize the MI waveguide in the field of wireless communications, which can greatly enlarge the communication range in many challenging environments. It should be noted that we adopt a theoretical propagation model that is similar to that shown in [96]. This model has been validated by experiments in [97].

The deployment problem in the MI-based wireless networks is related to the topology control problem in the traditional wireless networks. In [98], a robust and energy-efficient wireless network is constructed by adjusting the transmission power of each wireless node. In [99], a topology control strategy is proposed to guarantee connectivity in a wireless network, while the interference is minimized. In [100], an energy-aware topology control algorithm for a hierarchical sensor network is proposed. In [101], a comprehensive survey on the topology control protocols that utilize the geometric structures and virtual backbones is provided. In [102], the virtual force algorithm is introduced for wireless sensor networks. In [103], the Voronoi diagram-based mobile sensor deployment protocols are proposed to discover and fill the coverage holes. All the above works focus on the wireless networks that use traditional EM wave-based techniques. None of the existing works can address the deployment problems in the MI-based networks; they have fundamental differences: (1) the nodes in traditional wireless networks can directly connect to each other, while the nodes in the MI-based network can only be connected through the MI waveguides; and (2) the nodes in traditional wireless networks have a disk-shaped communication range, while the MI waveguide in the MI-based network has a much more complicated communication range, as shown in Fig. 6.7. In this section, we investigate the optimal MI waveguide deployment strategy to construct a low-cost and fully connected MI-based network that is robust to devise failures and coil displacements in challenging environments.

6.2.3 Optimal MI Waveguide Deployment Algorithm

In this section, the optimal MI waveguide deployment strategy is developed for the MI-based wireless networks in challenging environments. Specifically, we start with the deployment strategy in the basic 1-D network and then focus on the optimal deployment in the more general 2-D network. Since the 2-D network is constituted by multiple 1-D links, the analysis results of the 1-D deployment strategy lay the foundation of the MI waveguide deployment strategy in 2-D MI-based networks. It should be noted that this section focuses on the deployment strategy to construct a fully connected MI network. The interferences between multiple MI transceivers are assumed to be effectively eliminated by some well-designed MAC/scheduling protocols, which we have analyzed in another paper [104].

MI Waveguide Deployment in 1-D MI-Based Networks

In 1-D networks, wireless nodes are located along a polygonal line. This type of network topology is suitable for many applications where wireless nodes are placed along a chain, such as the underground pipeline monitoring system and the border patrol system. One-dimensional networks can be divided into multiple links connecting adjacent nodes. The goal of MI waveguide deployment in a 1-D network is to use as few relay coils as possible to connect the two nodes in each link. A certain level of robustness to the node failure and coil displacement is also required. The optimal number of relay coils for each link is determined by the path loss of the link, which is the function of the length of the link, the coil parameters, and the expected network reliability. In this subsection, we first give the path loss of the MI waveguide communications and then use the derived path loss to calculate the optimal relay coil number of each 1-D link.

In [90, 91], the path loss of the ideally deployed MI waveguide is provided. In this subsection, we extend the results to cover the scenarios when the relay coils constituting the MI waveguides are randomly displaced from the ideal positions. Assuming that the length of a link is d, an MI waveguide with $n-1$ relay coils is deployed along the link. The angle frequency of the signal is ω. In the ideal case, the relay coils are placed horizontally in a planar line, as shown in Fig. 6.8. This MI waveguide structure guarantees the omnidirectional coverage of each relay coil, which eases the deployment of

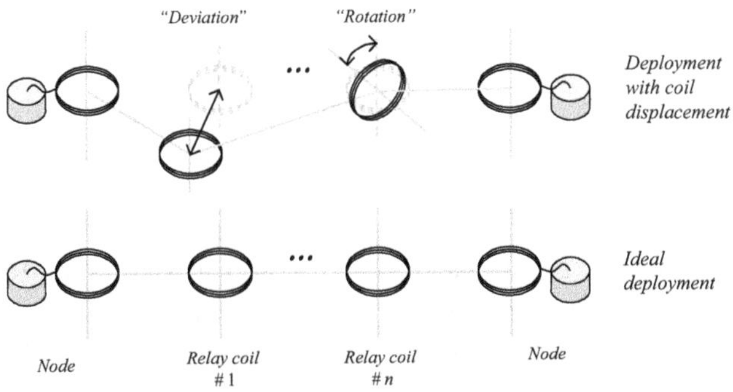

Figure 6.8 Illustration of the possible relay coil displacements [92]

the coils in challenging environments. The ideal intervals r_i $(i = 1,2,...,n)$ between every two adjacent relay coils are all the same and $r_i = d/n$.

However, in the practical operations, the position of each coil in the MI waveguide may deviate from the ideal positions due to the perturbations in challenging environments. As shown in Fig. 6.8, the horizontal and vertical deviations as well as the rotation of each coil may cause the changes in the coil intervals r_i and in the axial direction of each coil. As a result, the mutual inductions M_i $(i = 1,2,...,n)$ between the adjacent relay coils may be different:

$$M_i \simeq \mu\pi N^2 \frac{a^4}{4r_i^3} \cdot (2\sin\theta_{t_i}\sin\theta_{r_i} + \cos\theta_{t_i}\cos\theta_{r_i}), \tag{6.21}$$

where μ is the medium permeability, N is the number of turns of the coils, a is the coil radius, θ_t and θ_r are the angles between the coil radial direction and the line connecting the coil centers.

Then based on the results provided in [90, 91], the path loss (in dB) of the MI waveguide under the impact of coil displacement can be expressed as

$$L_{\rm MI}(d,n,\omega) \simeq 6.02 + 20\text{In}\left[\frac{Z}{\omega M_n} \cdot \zeta(\tfrac{Z}{\omega M_i}, n-1)\right], \tag{6.22}$$

where Z is the self-impedance of one relay coil, ω is the operating angular frequency, and $\zeta(\frac{Z}{\omega M_i}, n)$ is a polynomial of $\frac{Z}{\omega M_i}$ $(i = 1,2,...,n)$. The self-impedance of a coil Z is designed to be resonant at the center frequency ω_0. When $\omega = \omega_0$, Z becomes pure resistance R, which is the coil wire resistance. The polynomial $\zeta(\frac{R}{\omega M_i}, n)$ can be developed as

$$\begin{aligned}
\zeta(\tfrac{R}{\omega M_i}, 0) &= 1,\\
\zeta(\tfrac{R}{\omega M_i}, 1) &= \frac{R}{\omega M_1} + \frac{\omega M_1}{R},\\
&\vdots\\
\zeta(\tfrac{R}{\omega M_i}, n) &= \frac{R}{\omega M_n} \cdot \zeta(\tfrac{R}{\omega M_i}, n-1) + \frac{M_n}{M_{n-1}} \cdot \zeta(\tfrac{R}{\omega M_i}, n-2).
\end{aligned} \tag{6.23}$$

Substituting (6.40) into (6.39) yields the path loss of the MI waveguide under the impact of coil displacement. It should be noted that the mutual induction M_i is the random variable due to the random coil displacement. Consequently, the path loss is also a random variable.

To minimize the deployment cost and guarantee the network working properly most of the time, the MI waveguide should use the minimum number of relay coils to maintain a very low probability that its path loss is smaller than a threshold. To derive such an optimal number, the complicated link path losses given in (6.40) and (6.39) need to be simplified first. In (6.40), the value of $\frac{R}{\omega M_i}$ is large enough, especially in the case that the coil interval r_i is set as large as possible in order to reduce the coil number. Since $\zeta(\frac{R}{\omega M_i}, n)$ is a polynomial of $\frac{R}{\omega M_i}$, the highest order variable in the polynomial has the most influence. Therefore, (6.40) can be approximately expressed as

$$L_{\text{MI}}(d,n,\omega) \simeq 6.02 + 20\text{In}\left(\prod_{i=1}^{n} \frac{R}{\omega M_i}\right). \tag{6.24}$$

Assuming that the transmission power of each node is p_t, and the minimum received power for correct demodulation is p_{th} (both in dBm). We expect that the MI waveguide link has the outage probability (i.e., the received power is smaller than the threshold) smaller than P_{outage}:

$$P\left(p_t - L_{\text{MI}}(d,n,\omega) < p_{\text{th}}\right) < P_{\text{outage}}. \tag{6.25}$$

If the optimal relay coil number is denoted as n_{opt}, according to the previous discussion and substituting (6.24) into (6.25), we have

$$P\left(p_t - 6.02 - 20\text{In}\left(\prod_{i=1}^{n_{\text{opt}}} \frac{Z}{\omega M_i}\right) < p_{\text{th}}\right) = P_{\text{outage}}. \tag{6.26}$$

Therefore,

$$P\left(\prod_{i=1}^{n_{\text{opt}}} M_i < \frac{(\frac{R}{\omega})^{n_{\text{opt}}}}{10^{\frac{p_t-6.02-p_{\text{th}}}{20}}}\right) = P_{\text{outage}}. \tag{6.27}$$

If all relay coils are fixed at the ideal positions, the coil interval is fixed at $r_i = d/n_{\text{opt}}$. Hence, all the mutual inductions M_i are the same, and the path loss is a determined value. Therefore, in the case of the ideal deployment, the optimal number of relay coils for a link should fulfill the following equation:

$$n_{\text{opt}}(d,B) = \arg\min_{n}\{P_t - L_{\text{MI}}(d,n,\omega_0 + 0.5B) \geq P_{\text{th}}\}. \tag{6.28}$$

Due to the random coil deviation, the mutual inductions M_i of all links are random variables. The distribution of M_i is influenced by many perturbations determined by the geographic characteristics and specific applications. Hence, the exact distribution functions of M_i are very complicated and different from case to case. In this section, we approximately model the distribution of M_i to shed light on how the coil displacement influences the network reliability and the optimal coil number. We model the M_i of each link as the independent and identical random variables that are uniformly distributed in the interval $[M_{\text{min}}, M_{\text{max}}]$. The values of M_{min} and M_{max} are determined by the intensity of the coil displacement. According to (6.41), the coil displacement may change the coil interval r_i and the directions θ_t and θ_r. The term $(2\sin\theta_{t_i}\sin\theta_{r_i} + \cos\theta_{t_i}\cos\theta_{r_i})$ ranges from 0 to 2 and is 1 when coils are ideally placed. The term r_i ranges from d/n to infinity and is d/n when coils are ideally placed. Therefore, if the coil displacement intensity is $a\%$, M_i is uniformly distributed in the interval $[M_{\min}(d,n,b\%), M_{\max}(d,n,b\%)]$, where

$$M_{\min}(d,n,b\%) = \mu\pi N^2 \frac{a^4}{4\left[\frac{d}{n}(1+b\%)\right]^3}\cdot(1-b\%)$$
$$M_{\max}(d,n,b\%) = \mu\pi N^2 \frac{a^4}{4\left(\frac{d}{n}\right)^3}\cdot(1+b\%). \tag{6.29}$$

Since M_i are uniform, independently, and identically distributed random variables, the PDF of $X = \prod_{i=1}^{n} M_i$ can be expressed as [105]

$$f_X(x) = \begin{cases} \sum_{j=0}^{n-k} \frac{(-1)^j}{(M_{\max}-M_{\min})^n(n-1)!}\binom{n}{j}\left(\ln\frac{M_{\max}{}^{n-j}M_{\min}{}^{j}}{x}\right)^{n-1}, \\ \quad \text{if } M_{\min}{}^{n-k+1}M_{\max}{}^{k-1} \le x \le M_{\min}{}^{n-k}M_{\max}{}^{k},\ k=1,\ldots,n \\ 0, \quad \text{else} \end{cases} \tag{6.30}$$

Equation (6.27) can be developed as

$$P\left(\prod_{i=1}^{n_{\text{opt}}} M_i < \frac{\left(\frac{R}{\omega}\right)^{n_{\text{opt}}}}{10^{\frac{p_t-6.02-p_{\text{th}}}{20}}}\right) = \int_{M_{\min}{}^n}^{\frac{(\frac{R}{\omega})^{n_{\text{opt}}}}{10^{\frac{p_t-6.02-p_{\text{th}}}{20}}}} f_X(x)\,dx = P_{\text{outage}}. \tag{6.31}$$

By substituting (6.30) and (6.29) into (6.31), the optimal relay coil number can be derived by numerically solving the resulted transcendental equation. The optimal number of relay coils is the function of the link length d, the outage probability P_{outage}, the displacement intensity $b\%$, the operating frequency ω, and the coil parameters, such as the coil size, number of turns, and wire resistance. Since usually the outage probability, the operating frequency, and the coil parameters are fixed, the only variables in deployment design are the link length and the displacement intensity.

By using the parameters of the MI waveguide developed in [90], we can numerically analyze the optimal number of relay coils with different link length and displacement intensity. In the numerical analysis, the transmission power is 2.5 mW (4 dBm). The power threshold for correct reception is −90 dBm. The maximum tolerable outage probability is 5%. The operating frequency is 10 MHz. The relay coils have the same radius of 0.15 m, and the number of turns is 20. The wire resistance of unit length is 0.01 Ω/m. The MI-based wireless network is deployed in the soil medium. The permeability is a constant and is similar to that of the air, since most soil in the nature does not contain magnetite. Therefore, $\mu = 4\pi \times 10^{-7}$ H/m. The soil moisture and the soil composition do not affect the MI communication as discussed previously.

In Fig. 6.9, the optimal number of relay coils for one link in a 1-D MI-based network with different coil displacement intensities are shown as a step function of the link length. Since the dashed line connecting each step constitutes a convex function, the optimal relay coil number increases faster than the link length. This phenomenon is

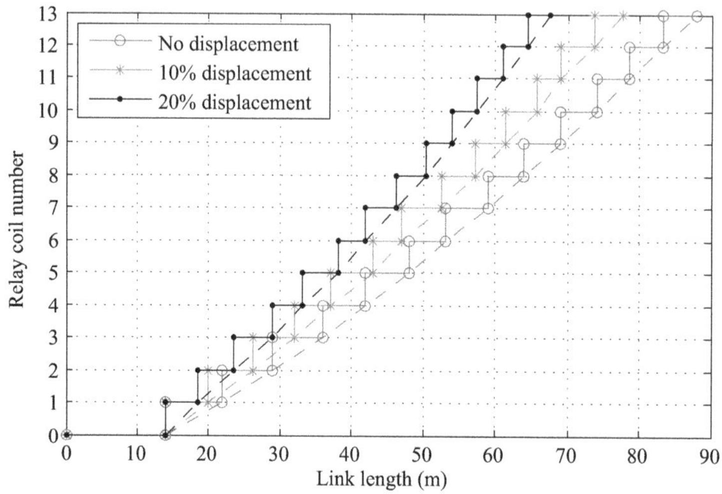

Figure 6.9 Optimal relay coil number with different link length and displacement intensity [92]

due to the fact that the coils relay the signal in a passive way, and no extra power is added at each relay coil. If coil displacements exist, a dramatically larger coil number is required to keep the outage probability low. As the displacement intensity increases, the required coil number increases even faster. Moreover, the coil displacement has more obvious impact when the link length increases. Hence, both the link length and the coil displacement should be kept below a threshold.

Besides the coil displacements, it should be noted that the 1-D network is very sensitive to the node failure; hence, the network will be partitioned into disconnected parts if any node dies. To enhance the robustness of node failure in a 1-D network, the only solution is to use the MI waveguide to connect multi-hop neighbors. The same equations can be used to calculate the optimal coil number. The only difference is that the parameter d is changed to the sum of multiple links.

Topology Model of the 2-D MI-Based Networks

In most applications, the network is deployed in a 2-D plane, such as the intelligent irrigation system, the earthquake and landslide forecast system, and the mine disaster prevention and rescue system, among others. In the remainder of this section, we investigate the deployment strategies of MI waveguides to connect the nodes in a 2-D network. Compared with 1-D networks, deployment of nodes in 2-D networks is much more complicated: (1) In 1-D networks, the route connecting the wireless nodes, i.e., the path connecting the nodes along the 1-D polygonal line, is determined. In contrast, the optimal route to connect all the wireless nodes in the 2-D plane needs to be found out. (2) It is possible in a 2-D network that some common relay coils can be shared by multiple links so that the total number of relay coils can be reduced.

The positions where the MI waveguides are deployed are first determined by the distribution of the nodes, which can be divided into two categories: the random and the regular distribution. If the nodes can be placed at any desired positions, the regular distribution, such as the hexagonal tessellation, can be used due to the high efficiency and simplicity. Otherwise, if the node has to be placed at certain positions due to environmental constraints or application requirements, the node distribution is random. In the rest part of this section, we first investigate the optimal deployment in the general case, i.e., the 2-D MI-based network with random node distribution. Then we focus on the special case, i.e., the network with regular node distribution, to achieve more explicit results.

MI Waveguide Deployment in 2-D Networks with Random Node Distribution

As discussed in Section 6.2.3, the objective of the optimal deployment includes: (1) using as few as relay coils as possible and (2) constructing a network as robust as possible to the node failure and the coil displacement. In this subsection, we investigate the optimal deployment strategies for the 2-D MI-based networks with random node distribution. The positions of the wireless nodes in such 2-D networks are supposed to follow a homogeneous Poisson point process with the spatial density λ_{rand} (m^{-2}). We first propose a minimum spanning tree (MST) algorithm that can minimize the total relay coil number but is not robust at all. The constructed network is not efficient in power consumption and routing, either. To address the problem, we propose a Voronoi diagram-based route that achieves optimal balance between system robustness and cost. The constructed network topology is also geometrical and power spanner. Finally, we provide a Fermat point-based method to further reduce the deployment cost without losing the spanner property and robustness.

If robustness is not considered, the most important goal is to connect all the nodes with minimum number of relay coils. To this end, the MST [106] can be used to establish a connected MI-based wireless network, where the weight of each edge is the optimal relay coil number.

According to the properties of the MST, the MST algorithm can construct a connected MI-based network with minimum number of relay coils. However, the failure of any one node disconnects the whole network. Moreover, the network is disconnected if the outage caused by the coil displacement happens in any one MI waveguide. In addition, the spanning tree topology may cause high congestions and end-to-end power consumption. Therefore, although the MST algorithm achieves the minimum deployment cost, it is not the optimal deployment strategy.

To find the optimal deployment strategy, we first identify the metrics to quantitatively characterize the robustness. The most accurate metric is the network outage probability, i.e., the probability that the data transmission from an arbitrary source to an arbitrary destination is failed due to the disconnected network. However, since the node distribution is highly random, it is impossible to derive the analytical expression of the network outage probability. Hence, we use a more straightforward network parameter, the average node degree (1-hop neighbor number), to characterize the network robustness. Since the node distribution follows a homogenous Poisson point process, the average node degree can, to a large extent, represent the network outage behavior. Then, the deployment objective becomes to find the optimal balance

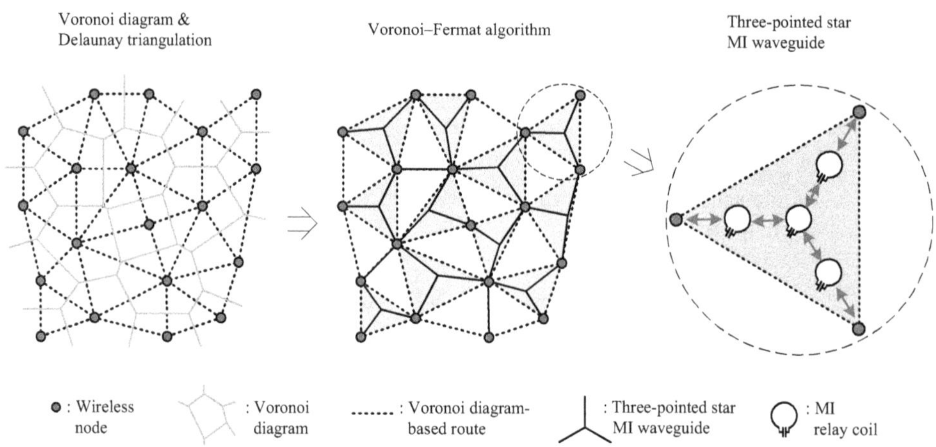

Figure 6.10 The MI waveguide deployment using the VF algorithm in the MI-based networks with random node distribution [92]

Algorithm 3: Voronoi–Fermat (VF) algorithm for MI waveguide deployment

---

Create the Voronoi diagram of the K wireless nodes, and derive K Voronoi cells $\mathbf{VC} = \{VC_1, VC_2, ..., VC_K\}$. Keep a subset $\mathbf{G}$ of $\mathbf{VC}$; $\mathbf{G}$ initially contains VC_1.
while *(Not all Voronoi cells are in* **G***)* **do**
  Find a Voronoi cell VC_x in $\mathbf{G}$ that has the neighbor Voronoi cells $\{VC_x^1, VC_x^2, ..., VC_x^j\}$ which are not in $\mathbf{G}$.
  Connect the adjacent wireless nodes in $\{VC_x^1, VC_x^2, ..., VC_x^j\}$ and VC_x, and derive the non-overlapped triangle cells $\{Tr_1, Tr_2, ..., Tr_{j-1}\}$.
  if *(j is odd)* **then**
    In triangle cells $Tr_1, Tr_2, Tr_4, ..., Tr_{j-1}$, deploy the MI waveguide along the three lines connecting the vertexes and the Fermat point
  else
    In triangle cells $Tr_1, Tr_3, Tr_5, ..., Tr_{j-1}$, deploy the MI waveguide along the three lines connecting the vertexes and the Fermat point.
    Add $\{VC_x^1, VC_x^2, ..., VC_x^j\}$ to $\mathbf{G}$.
  end
end

---

between the small coil number and the large node degree. The constructed network should also be efficient in reducing congestions and the power consumption.

To this end, we propose the Voronoi diagram-based route, along which the MI waveguides are deployed. As shown in Fig. 6.10, the Voronoi diagram of the wireless nodes partitions the whole area into Voronoi cells. Each cell contains one node. Any point in one Voronoi cell is closer to the node in this cell than to any other nodes. The Voronoi diagram-based route is constructed by connecting the nodes in adjacent cells, as shown in Fig. 6.10. Then the MI waveguides are deployed along this Voronoi diagram-based route. If every transmission pair has to use a unique MI waveguide, the Voronoi diagram-based route is optimal due to the following properties.

Proposition 1 The average node degree of the MI-based network constructed by the Voronoi diagram-based route is approximately 6.

The proof of proposition 1 is given in Appendix A. According to this proposition, the Voronoi diagram-based route achieves a much higher average node degree ($E(m) = 6$) than the MST algorithm ($E(m) \simeq 2$). Since the network robustness increases exponentially with the node degree, the Voronoi diagram-based route significantly enhances the robustness, which is also clearly shown in the numerical studies in Section 6.2.4.

Next, we quantitatively exam the balance between the deployment cost and network robustness. We define a new matric, the *cost-robust factor* $\rho_{\text{cost-robust}}$, to characterize such a balance, which is the ratio of the average coil number per node to the average node degree in the constructed network. The smaller the $\rho_{\text{cost-robust}}$, the higher the cost-robust efficiency. Now, we have the following proposition.

Proposition 2 The average cost-robust factor of the network with the Voronoi diagram-based route is of the same order as that of the network constructed by the MST algorithm.

The proof of proposition 2 is given in Appendix B. Propositions 1 and 2 indicate that the Voronoi diagram-based route achieves much higher network robustness than the MST algorithm by introducing a reasonable number of relay coils.

Besides the network robustness, congestions and power consumption should also be considered. Hence, we introduce the following propositions.

Proposition 3 The MI-based network constructed according to the Voronoi diagram-based route is planar and geometrical spanner.

As mentioned before in Proposition 1 and 2, the Voronoi diagram-based route forms a Delaunay triangulation, which has been proved to be planar and geometrical spanner [101]. Therefore, Proposition 3 is proved.

The geometrical spanner property indicates that the shortest path connecting any two nodes in the constructed MI-based network is not longer than k ($k < \infty$) times of the Euclidean distance between them. This property guarantees that the data between any source and destination nodes can be transmitted through a relatively short path, which is necessary for the power efficiency, quality of service, and reliability of MI-based wireless networks. Moreover, the planar property of the resulted network is preferred for most geographic routing protocols, which is favorable for resource-limited devices [101]. In the traditional wireless networks, it has been proved that the network with a Delaunay triangulation topology is also a power spanner, i.e., the total power consumption along the shortest path connecting any two nodes in such a network is not much larger than the power consumption of the direct transmission between them. In wireless networks using MI waveguides, due to the complete different channel model, the power spanner property needs to be reexamined. Therefore, we have the following proposition.

Proposition 4 The MI-based network with the Voronoi diagram-based route is also a power spanner.

The proof of proposition 4 is given in Appendix C. According to Propositions 3 and 4, the MI-based networks with the Voronoi diagram-based route achieve the same routing, congestion, and power consumption performance as the network with the complete graph topology. To guarantee the spanner properties, no edge on the route should be deleted. Similarly, to guarantee the planar properties, no edge can be added, either.

The Voronoi diagram-based route significantly enhances the network robustness by adding a reasonable number of relay coils. It also improves the routing and energy performance due to the planar and spanner properties. If we assume that each link has to use a unique MI waveguide, the Voronoi diagram-based route is optimal for the MI waveguide deployment. If this constraint can be released, i.e., multiple links can share the same sets of relay coils, the relay coil number can be further reduced.

Our objective is to reduce the coil number but keep the same network topology of the Voronoi diagram-based route. To this end, we propose the Fermat point-based improvement: The three MI waveguides along the edges in each Delaunay triangle cell can be replaced by one MI waveguide with the shape of a three-pointed star, as shown in Fig. 6.10. The nodes on all the three vertices can use the same waveguide to communicate with each other directly. The center of the three-pointed star is located at the Fermat point of the triangle. Since the total distance from the three vertices to the Fermat point is the minimum possible [107], the three-pointed star MI waveguide centered at the Fermat point consumes the minimum number of coils to connect the three nodes.

To construct a network with the same topology as the original Voronoi diagram-based route, the three-pointed star-shaped MI waveguides only need to be deployed in every other Delaunay triangle cell, as shown in Fig. 6.10. By this way, the relay coil number required to construct the Voronoi diagram-based route is significantly reduced. Meanwhile, the network topology and the robustness to node failure remain the same since all the original links are not affected. It should be noted that the robustness to coil displacement is slightly weakened since every three links share the same MI waveguide. However, the impact of coil displacement can be limited to a very small extent according to the optimal deployment strategy in Section 6.2.3. Hence, the total network robustness can be viewed to be the same as the original Voronoi diagram-based route.

To sum up, the deployment strategy using both the Voronoi diagram-based route and the Fermat point-based improvement achieves high network robustness as well as low congestion and power consumption with minimum relay coil number. We denote this optimal deployment as the Voronoi–Fermat (VF) algorithm. The detailed procedure of the VF algorithm is described in Algorithm 3 in the book.

MI Waveguide Deployment in 2-D Networks with Hexagonal Tessellation Topology

On the basis of the deployment analysis in the general scenario with random node distribution, in this subsection, we focus on the special scenario where the nodes are distributed with regular patterns, especially the hexagonal tessellation. More explicit expressions can be derived due to the regular pattern. Hexagonal tessellations have been widely used for the traditional wireless network topologies, such as the base station placement of the cellular networks due to the efficiency in coverage. In the following analysis, we assume that the node density of the MI-based network with hexagonal tessellation topology is λ_{hex} (m^{-2}).

Similar to the network with random node distribution, the MST algorithm can be used to minimize the relay coil number without considering the network robustness. The average neighbor number of the resulted network is approximately 2, which is the same as in the random node distribution scenario.

Next, we calculate the cost-robust factor $\rho_{\text{cost-robust}}$. If the total number of the wireless node is K, the total number of links in the MST is $K-1$. The edges of the hexagonal tessellation have the same length d^{hex}, which is determined by the node density λ_{hex}:

$$d^{\text{hex}} = 2 \cdot 3^{-1/4} \cdot \lambda_{\text{hex}}^{-1/2}. \tag{6.32}$$

By utilizing the strategy given in Section 6.2.3, the optimal number of relay coils n_{opt} for each link can be calculated as a function of the link length d^{hex} and the tolerable outage probability $p_{\text{coil}}^{\text{out}}$ for a single MI waveguide. Then the total number of the relay coils to connect the K wireless nodes based on the MST algorithm is

$$\text{Num}_{\text{mst}}^{\text{hex}} = (K-1) \cdot n_{\text{opt}} \left(2 \cdot 3^{-1/4} \cdot \lambda_{\text{hex}}^{-1/2}, p_{\text{coil}}^{\text{out}}\right). \tag{6.33}$$

According to the cost-robust factor equation, we have

$$\rho_{\text{cost-robust}}{}^{\text{hex,mst}} = \frac{\text{Num}_{\text{mst}}^{\text{hex}}}{2 \cdot (K-1)} = \frac{1}{2} \cdot n_{\text{opt}} \left(2 \cdot 3^{-1/4} \cdot \lambda_{\text{hex}}^{-1/2}, p_{\text{coil}}^{\text{out}}\right). \tag{6.34}$$

Assuming that each wireless node has an independent and identical probability of failure $p_{\text{node}}^{\text{fail}}$, while the tolerable outage probability for each MI waveguide is $p_{\text{coil}}^{\text{out}}$. Then we can approximately evaluate the outage probability of the entire network, i.e., the probability that the data transmission from an arbitrary source to an arbitrary destination is failed due to the disconnected network. For an h-hop transmission, the outage probability $p_{\text{hex}}^{\text{out}}(h)$ is

$$p_{\text{hex,mst}}^{\text{out}}(h) = 1 - \left(1 - p_{\text{node}}^{\text{fail}}\right)^h \cdot \left(1 - p_{\text{coil}}^{\text{out}}\right)^h. \tag{6.35}$$

According to (6.34), the outage probability for any transmission is at least $p_{\text{node}}^{\text{fail}} + p_{\text{coil}}^{\text{out}} - p_{\text{node}}^{\text{fail}} \cdot p_{\text{coil}}^{\text{out}}$ for a single hop and can easily approach 1 when the hop number increases. Since the average transmission hop is large in the MST topology, the average outage probability of the network constructed by the MST algorithm is extremely large.

The VF algorithm in the network with hexagonal tessellation topology is much more straightforward than in the random distributed network. The links of the Voronoi diagram-based route are exactly the edges of the hexagonal tessellation, while the Fermat point of each equilateral triangle cell is located at the centroid. In the constructed MI-based network, each wireless node has exactly six neighbors, which is much more than the MST algorithm (two neighbors). The cost-robust factor $\rho_{\text{cost-robust}}{}^{\text{hex}}_{\text{VF}}$ can also be derived in the same way as in the MST algorithm. Therefore,

$$\rho_{\text{cost-robust}}{}^{\text{hex, VF}} = \frac{\text{Num}^{\text{hex}}_{\text{VF}}}{2 \cdot 3\text{K}} = \frac{1}{4} \cdot n_{\text{opt}} \left(4 \cdot 3^{-3/4} \cdot \lambda_{\text{hex}}^{-1/2}, p_{\text{coil}}^{\text{out}}\right). \tag{6.36}$$

The outage probability of the network constructed by the VF algorithm can be approximated by summing up the isolated probabilities of the transmitter and the receiver, i.e.,

$$p_{\text{hex, VF}}^{\text{outage}} \simeq 2 \cdot \left(p_{\text{node}}^{\text{fail}} + p_{\text{coil}}^{\text{out}} - p_{\text{node}}^{\text{fail}} \cdot p_{\text{coil}}^{\text{out}}\right)^{6}. \tag{6.37}$$

It should be noted that this outage probability is applicable for any source–destination pairs disregarding how far they are apart from each other. By comparing the results given in (6.34)–(6.37), we can clearly see that the VF algorithm achieves much larger neighbor number (node degree), lower cost-robust factor (i.e., higher cost-robust efficiency), and dramatically lower outage probability than the MST algorithm.

6.2.4 Numerical Evaluation

In this section, we numerically evaluate the required relay coil number and the network robustness of the MST algorithm and the VF algorithm in MI-based networks with both random node distribution and the hexagonal tessellation topology. In the following simulations, 100 wireless nodes are deployed in a square area according to the hexagonal tessellation topology or the homogenous Poisson point process. The size of the square area is determined by the node density. The relay coil number for each MI waveguide is determined by (6.29)–(6.31), where the MI waveguide parameters are the same as the parameters used in Section 6.2.3; the displacement intensity of the relay coils is set to be 10%; and the tolerable outage probability for each MI waveguide is set to be 5%.

Figure 6.11 shows the deployment results of the MST algorithm and the VF algorithm. The network constructed by the Voronoi diagram-based route without the Fermat improvement is also listed for comparison. Note that the deployment result in the hexagonal tessellation topology is omitted due to its simplicity and the page limit. As shown in Fig. 6.11(a), the network constructed by the MST algorithm has the least number of links. Consequently, the relay coil number is small, but the network is not robust. The failure of any nodes or the outage of any MI waveguide will result in a disconnected network. In contrast, as shown in Figs. 6.11(b) and (c), the networks constructed by the VF algorithm and Voronoi diagram-based route have the same network topology, which have much more links connecting adjacent wireless nodes. Therefore, the resulted networks are much more robust to node failures but may consume more relay coils.

In Figs. 6.12(a) and (b), the total relay coil numbers of the deployment algorithms are given as a function of the node density in networks with hexagonal tessellation topology and random node distribution, respectively. Compared with the MST algorithm, the relay coil number required by the VF algorithm is almost the same in the

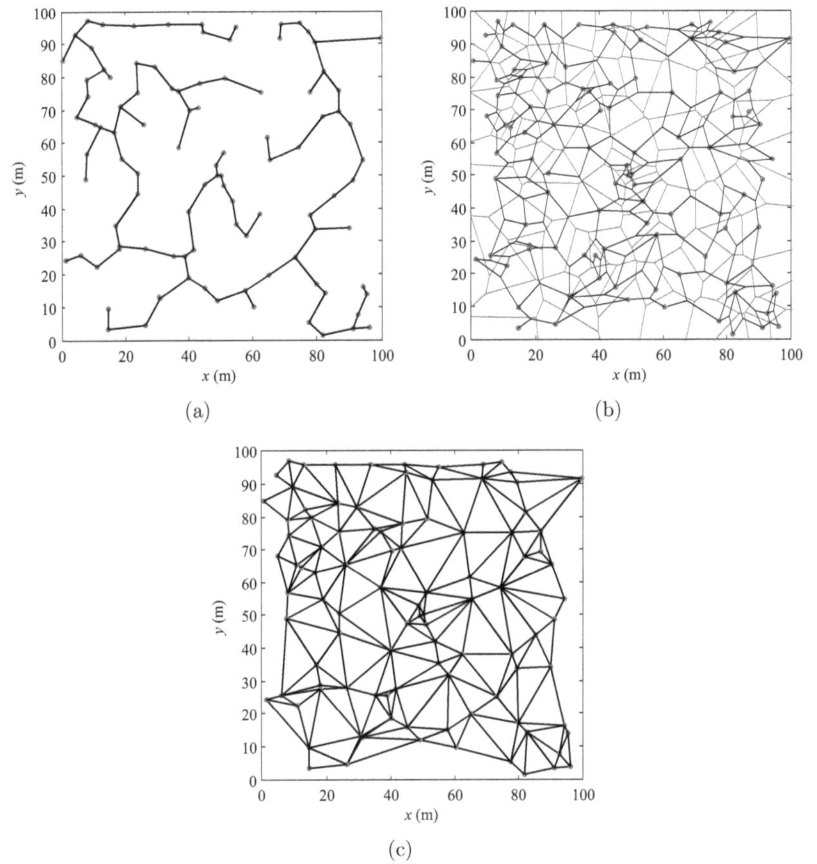

Figure 6.11 The deployment results of (**a**) the MST algorithm, (**b**) the VF algorithm, and (**c**) the VF strategy without the Fermat improvement (The red dots are the wireless nodes, the black lines represent the MI waveguides, and the blue cells are the Voronoi diagrams. Hundred wireless nodes are uniformly distributed with a spatial intensity $\lambda_{\text{rand}} = 0.01\ \text{m}^{-2}$)

networks with hexagonal tessellation topology and slightly higher in the networks with random node distribution. In both network topologies, the coil number required by the Voronoi diagram-based route is much larger. Next, to see whether the more relay coils bring higher robustness or not, we check the robustness of the networks constructed by the VF algorithm and the Voronoi diagram-based route.

In Fig. 6.13, we exam the network outage probability, i.e., the probability that the data transmission of an arbitrary source–destination pair is failed due to network connectivity. The network outage probabilities of the networks constructed by different deployment algorithms are plotted as the function of the failure probability of any single wireless node. It shows that the outage probabilities of the MST algorithm in both the hexagonal tessellation topology and the random topology are similar and close to 1 in such large-scale networks (100 nodes). In contrast, the network built by the VF algorithm has extremely low outage probability in large-scale networks, especially in the hexagonal tessellation topology network since every node in such networks is

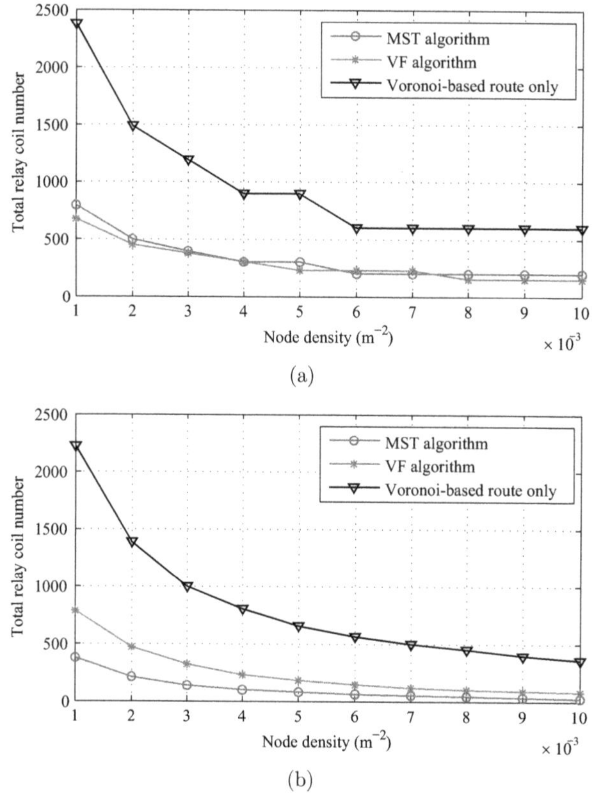

Figure 6.12 The number of relay coils to construct an MI-based network with (**a**) hexagonal tessellation topology and (**b**) random topology [92]

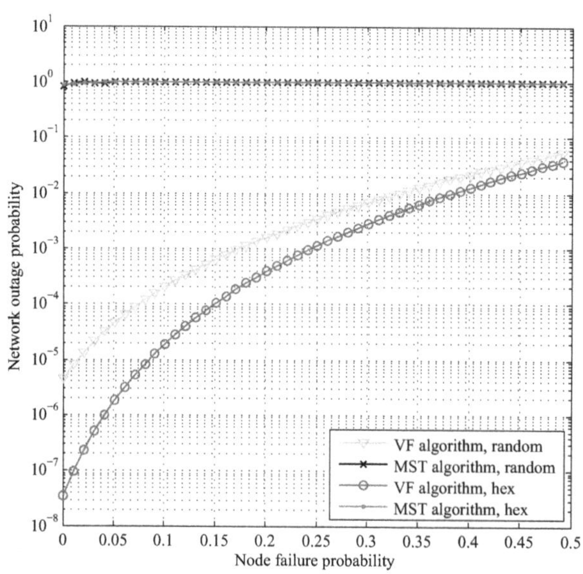

Figure 6.13 The network outage probability [92]

guaranteed to have six neighbors. It should be noted that the outage behavior of the Voronoi diagram-based route is the same as the VF algorithm due to the same link topology.

According to the theoretical and numerical results provided in this section and Section 6.2.3, the MST algorithm can construct an MI-based network with the minimum number of relay coils, but the robustness to node failure and coil displacement is extremely poor. Meanwhile, the VF algorithm can greatly enhance the network robustness by just adding a small number of relay coils (even no more coils are needed in the hexagonal tessellation topology).

6.3 Network Capacity

6.3.1 Introduction

By bringing the wireless sensor networks into the underground soil environment, the WUSNs enable a wide variety of novel and important applications, such as intelligent irrigation, mine disaster prevention and rescue, concealed border patrol, in situ sensing for oil recovery, and underground infrastructure monitoring, among others [108]. However, it is difficult to establish efficient wireless links among underground sensor nodes, since the traditional wireless communication techniques based on EM waves encounter two major problems in the soil medium: (1) extremely small communication ranges (<5 m) due to high path loss and (2) highly dynamic channel conditions caused by the time-varying soil properties such as soil moisture [109].

MI-based WUSNs use the novel MI waveguide technique [90, 95, 96] to establish efficient wireless communications in the underground soil medium. As shown in Fig. 6.14, the underground sensor nodes in MI-based WUSNs are wirelessly connected by MI waveguides consisting of multiple MI relay coils. Wireless communications are accomplished by the consecutive MI between the adjacent MI relay coils. The MI waveguide technique is favorable in WUSNs due to the following five advantages.

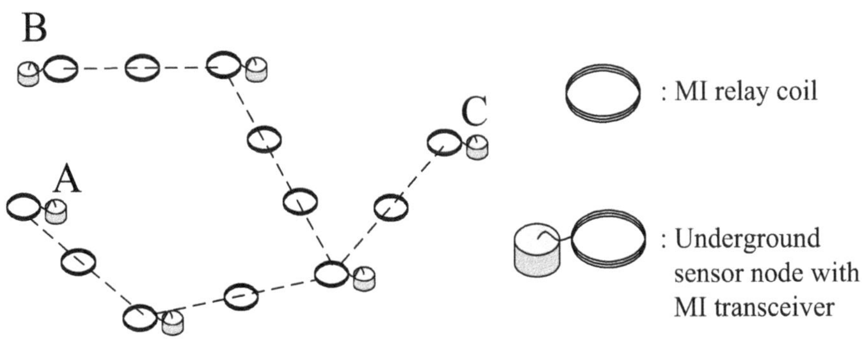

Figure 6.14 The network geometric structure of the MI-based WUSNs [111]

- The communication range between two underground sensor nodes is up to 100 m. Hence, a fully connected network can be achieved without very high sensor density.
- The MI channel conditions remain constant in most soil medium since the attenuation rate of magnetic fields does not change in most types of transmission media.
- The MI relay coils do not consume extra energy, and the unit cost is negligible.
- The relay coils are easy to deploy and do not need regular maintenance. The relay coils only need to be deployed every 6–12 m and can be flexibly deployed at any depth.
- The lifetime of MI-based WUSNs can be greatly prolonged since the underground sensor nodes equipped with MI transceivers can be recharged by aboveground devices using the inductive charging technique [110].

Despite numerous advantages, the capacity and reliability of the MI-based WUSNs are the primary concerns. On the one hand, the bandwidth of the MI waveguide technique is much smaller than the terrestrial EM wave-based techniques since the MI relay coils have to work at the resonant frequency to maintain low path loss. Although the low duty circle WUSN applications do not require a very high data rate, large-scale WUSNs can still generate a significant amount of traffic that need to be efficiently delivered on the bandwidth-limited MI channels via multi-hop wireless paths. Hence, a comprehensive analysis of the channel and network capacities of MI-based WUSNs is necessary. On the other hand, since the communication success of the MI waveguide technique relies on multiple resonant MI relay coils, the functionality of the MI-based WUSNs depends not only the underground sensor nodes but also on all the MI relay coils. Therefore, the reliability of MI-based WUSNs on needs to be examined in harsh underground environments.

The theoretical study of the capacity and reliability in MI-based WUSNs is challenging and dramatically different from that of traditional wireless networks. First, due to the unique signal propagation solution, the bandwidth and received signal strength of the MI channel experience completely different attenuation law as the transmission distance increases. Consequently, the MI channel capacity has dramatically different characteristics from the traditional wireless channels. Second, since the MI signal can only propagate along the MI relay coils, the communication and interference ranges of a sensor node are no longer disks but can be arbitrary shapes depending on the relay coil deployment strategy. For example, in Fig. 6.14, node A and a physically nearby node B can transmit at the same time without interference since they are far apart along the routes connected by the MI waveguides. However, A will be interfered by the simultaneous transmissions from a physically far away node C since they can be close enough to each other through the relay coils. Hence, the network geometric structure of MI-based WUSNs in terms of connectivity and interference is significantly different from traditional wireless networks. Due to the unique channel capacity and the network geometric structure, MI-based WUSNs also have dramatically different network capacity. Third, in MI-based WUSNs, not only sensor failure but also many other incidents, such as relay coil missing and displacement, can affect the system's performance. Hence, MI-based WUSNs encounter more severe reliability challenges than the traditional wireless networks. To our best knowledge, the above problems have not been addressed by the research community so far.

In this section, we theoretically analyze the channel capacity, network capacity, and the reliability of MI-based WUSNs. In particular, we first develop a mathematical expression of the channel capacity in WUSNs when the information propagation relies on the consecutive magnetic induction. The effects of multiple system parameters, including relay coil density, operating frequency, relay coil size, wire resistance, and number of turns of each relay coil, are captured. On the basis of the analysis of MI channel capacity, we investigate the network capacity scaling laws of MI-based WUSNs under different relay coil deployment strategies. The achievable throughput of each node is derived when the number of sensor nodes and the number of MI relay coils in the network increase. Finally, we discuss the impact of different factors on the reliability of the channel capacity and the network capacity in MI-based WUSNs. The factors that we studied include the sensor node failure, relay coil missing, and direction and position deviation of the relay coils. The channel and network capacities as well as the corresponding reliabilities derived in this section provide principles and guidelines for the design and deployment of MI-based WUSNs.

6.3.2 Related Work

The definition of WUSNs is first proposed in [108]. We have developed the channel model of EM waves in the soil medium in [109], which clearly shows the major problems of the EM wave-based techniques in WUSNs. This theoretical model has been validated by the testbed developed in [112]. In [113], we analyzed the connectivity of the WUSNs using EM waves. It is shown that an extremely high density of underground sensor nodes is required to form a fully connected network due to the small underground communication range.

In [90], we introduce the MI waveguide to WUSNs, which can greatly enlarge the underground communication range without incurring very high cost. The MI waveguide technique is first developed in [95, 96]. It is shown that an array of loops can act as a waveguide, propagating a new form of wave known as the MI wave. The theoretical propagation model has been validated by experiments in [114]. The MI waveguide is originally designed as artificial delay lines and filters, dielectric mirrors, distributed Bragg reflectors, slow-wave structures in microwave tubes, coupled cavities in accelerators, etc. In [90], we first utilize the MI waveguide in the field of wireless communications, where we adopt the same theoretical analysis method as [96]. In [115], we propose three strategies to deploy MI relay coils among unconnected underground sensor nodes to construct a fully connected network. The MI waveguide channel model in [90] and the MI waveguide deployment strategies in [115] provide the preliminary knowledge of the capacity and reliability analysis in Section 6.3.

The capacity of traditional wireless networks using EM wave-based techniques has been intensively investigated by the research community. The network capacity scaling law of the traditional ad hoc networks is first studied in the seminal work [116].

The authors show that the achievable throughput of each randomly located node decreases at the speed of $1/\sqrt{n \log n}$ as the number of nodes n in the network increases. In recent years, this seminal work has been extended to different types of wireless networks, such as the cognitive radio networks [117], the multi-channel multi-radio ad hoc networks [118], the ad hoc network with infrastructure [118], and the ad hoc networks with the MIMO or directional antenna [119, 120], among others. However, all the above works focus on the wireless networks that use traditional EM wave-based propagation techniques. None of the existing works considers a wireless network that has the same channel characteristics or network geometric structure as in the MI-based WUSNs. Moreover, there is no existing work considering the impact of harsh environmental factors on the reliability of the channel and network capacities in MI-based WUSNs.

6.3.3 Channel Capacity in MI-Based WUSNs

In this section, the capacity of the channel between two underground sensor nodes in the MI-based WUSNs is theoretically investigated. We start from the classic channel capacity formula given by

$$C = B \cdot \log\left(1 + \frac{P_t \cdot L_p}{N}\right), \tag{6.38}$$

where C is the channel capacity, B is the bandwidth of the channel, P_t is the transmission power, L_p is the path loss, and N is the total noise and interference power. In the traditional EM wave-based wireless channel, the bandwidth B is a constant and previously determined by the specific applications, while the channel path loss L_p is an exponential function of the transmission distance. On the contrary, in MI-based WUSNs, the MI waveguide technique is used instead of the EM wave-based technique. Consequently, the bandwidth B is no longer a constant but varies as the transmission distance increases or other system parameters change. Moreover, the path loss L_p of the MI waveguide is no longer a simple exponential function due to the consecutive magnetic induction. We have provided the channel model for the MI waveguide in soil medium in [90]. This channel model can be validated by the experiments conducted in [114]. In the following part of this section, we further develop this channel model to derive a close-form expression of the MI channel capacity, which lays the foundation of the network capacity analysis in the next section.

Overview of the MI Waveguide Channel Model

Consider the MI channel between two sensor nodes that are d m apart from each other. The deployment density of the MI relay coil is represented by the interval r between two adjacent relay coils, i.e., the relay coils are deployed every r m. Therefore, the MI waveguide consists of the $k-1$ relay coils, $k = \lceil d/r \rceil$. If the frequency of the transmitting signal is f, the path loss of the MI waveguide can be calculated according to [90]

$$L_p(d,r,f) \simeq 0.25|\zeta\left(\tfrac{Z}{2\pi fM},\lceil\tfrac{d}{r}\rceil\right)|^{-2}, \tag{6.39}$$

where $\zeta(\frac{Z}{\omega M},\lceil\frac{d}{r}\rceil)$ is the $\lceil\frac{d}{r}\rceil$-order polynomial of $\frac{Z}{\omega M}$, M is the mutual induction between the adjacent coils, and Z is the self-impedance of one relay coil. The polynomial $\zeta(x,k)$ can be developed as

$$\begin{aligned} \zeta(x,1) &= x, \\ \zeta(x,2) &= x^2+1, \\ &\vdots \\ \zeta(x,k) &= x\cdot\zeta(x,k-1)+\zeta(x,k-2). \end{aligned} \tag{6.40}$$

The mutual induction M can be deduced by the magnetic potential of the magnetic dipole:

$$M \simeq \mu\pi N^2\frac{a^4}{4r^3}\cdot(2\sin\theta_t\sin\theta_r+\cos\theta_t\cos\theta_r), \tag{6.41}$$

where μ is the permeability of the soil medium, N is the number of turns of the wire on the coils, a is the radius of the coils, and θ_t and θ_r are the angles between the coil radial directions and the line connecting the two coil centers.

Capacity of the MI Waveguide Channel

To derive the closed-form formula of the MI channel capacity, the path loss and bandwidth of the MI waveguide need to be explicitly expressed, where some approximation is required.

The polynomial $\zeta(x,n)$ defined in (6.40) can be further developed as a closed-form formula:

$$\zeta(x,k) = \frac{p^{2k}+(-1)^{k-1}}{p^{k-1}\cdot(p^2+1)}\cdot x+\frac{p^{2k-2}+(-1)^{k-2}}{p^{k-2}\cdot(p^2+1)}, \tag{6.42}$$

where $x=\frac{Z}{2\pi fM}$, $k=\lceil\frac{d}{r}\rceil$, and p is a function of x given by

$$p=\frac{x+\sqrt{x^2+4}}{2}. \tag{6.43}$$

In this section, we let all the MI relay coils face up so that the coil axial direction is vertical ($\theta_t=\theta_r=0$), as shown in Fig. 6.14. Although the planar coil deployment is not the direction that maximizes the mutual induction, it guarantees that each relay coil has omnidirectional communication range. The planar placed coils are also easy to deploy and are stable during operations. Moreover, to maintain low path loss, in the MI waveguide, the self-impedance of a coil Z is designed to be resonant at the center frequency f_0 [90]. When $f=f_0$, Z becomes pure resistance R, which is the coil wire resistance. Consequently, if $\theta_t=\theta_r=0$ and $f=f_0$, the variable x in the polynomial $\zeta(x,k)$ in (6.40) can be further developed as

$$x=\frac{Z}{2\pi fM}=\frac{2R\cdot r^3}{\mu\pi^2 f_0N^2a^4}. \tag{6.44}$$

Since the self-impedance of a coil Z is pure resistance R under the resonant status, the value of the polynomial $\zeta(x,n)$ becomes real. Then the path loss of the planar MI waveguide working at the central frequency is given by

$$L_p(d,r,f_0) = 0.25\,\zeta^{-2}\left(\frac{2R\cdot r^3}{\mu\pi^2 f_0 N^2 a^4}, \left\lceil \frac{d}{r} \right\rceil\right) \tag{6.45}$$

$$= 0.25\left[\frac{p^{2\lceil\frac{d}{r}\rceil} + (-1)^{\lceil\frac{d}{r}\rceil-1}}{p^{\lceil\frac{d}{r}\rceil-1}\cdot(p^2+1)}\cdot x + \frac{p^{2\lceil\frac{d}{r}\rceil-2} + (-1)^{\lceil\frac{d}{r}\rceil-2}}{p^{\lceil\frac{d}{r}\rceil-2}\cdot(p^2+1)}\right]^{-2},$$

where p and x are defined in (6.43) and (6.44), respectively.

Since all the relay coils work at the resonant status, the bandwidth of the MI waveguide is very small and dramatically varies if the transmission distance or other parameters change. Recall that the self-impedance of one relay coil $Z = R + j2\pi fL + \frac{1}{j2\pi fC}$, where R is the wire resistance, L is the self-induction of each coil, and C is the loaded capacitor to guarantee resonance. The capacitor C is designed to fulfill $j2\pi fL + \frac{1}{j2\pi fC} = 0$ at the central frequency f_0. The self-induction of each coil can be derived in the same way as the mutual induction. Hence, L and C can be calculated by

$$L \simeq \frac{1}{2}\mu\pi N^2 a\,, \qquad C = \frac{2}{4\pi^2 f_0^2 N^2 \mu\pi a}\,. \tag{6.46}$$

As the operating frequency deviates from the central frequency, the coil self-impedance Z is no longer pure resistance, and the absolute value dramatically increases. Consequently, the MI waveguide path loss also increases dramatically, which is the reason causing the small channel bandwidth. In this chapter, we consider the 3-dB bandwidth B as the channel bandwidth. Specifically, the path loss at the frequency $f_0 + 0.5B$ is two times the path loss at the central frequency f_0, i.e., $L_p(d,r,f_0 + 0.5B) = 0.5L_p(d,r,f_0)$. It should be noted that there may be multiple frequencies where the path loss is doubled. In this case, the smallest B is the real 3-dB bandwidth. Substituting (6.39) into $L_p(d,r,f_0 + 0.5B) = 0.5L_p(d,r,f_0)$ yields

$$\frac{\left|\zeta\left(\frac{R+j2\pi(f_0+0.5B)L+\frac{1}{j2\pi(f_0+0.5B)C}}{2\pi fM}, \left\lceil\frac{d}{r}\right\rceil\right)\right|}{\zeta\left(\frac{R}{2\pi f_0 M}, \left\lceil\frac{d}{r}\right\rceil\right)} = \sqrt{2}\,. \tag{6.47}$$

An equivalent approximation of (6.47) is needed to derive the closed-form expression of the bandwidth. As previously mentioned, $\zeta(x,k)$ is the k-order polynomial of x. Since $x = \frac{Z}{2\pi fM}$ is relatively large especially when the relay coil density is small (which is favorable for the deployment), the highest-order variable in the polynomial has the most influence. Moreover, the bandwidth B is much smaller than the central frequency f_0. Therefore, (6.47) is approximately equivalent to the following equation:

$$\left|\frac{R + j2\pi(f_0+0.5B)L + \frac{1}{j2\pi(f_0+0.5B)C}}{R}\right|^{\lceil\frac{d}{r}\rceil} = \sqrt{2}\,. \tag{6.48}$$

By substituting (6.46) into (6.48), the closed-form expression of the MI waveguide bandwidth can be approximately given by

$$B \simeq \frac{R}{\mu\pi^2 a}\sqrt{2^{\lceil \frac{d}{r} \rceil^{-1}} - 1}\,. \tag{6.49}$$

Substituting (6.45) and (6.49) into (6.38) yields the closed-form formula of the MI waveguide channel capacity:

$$C \simeq \frac{R}{\mu\pi^2 a}\sqrt{2^{\lceil \frac{d}{r} \rceil^{-1}} - 1} \cdot \log\left[1 + \frac{P_t \cdot \zeta^{-2}\left(\frac{2R\cdot r^3}{\mu\pi^2 f_0 N^2 a^4}, \lceil \frac{d}{r} \rceil\right)}{4N}\right]. \tag{6.50}$$

Numerical Analysis

The channel capacity in MI-based WUSNs is numerically analyzed with different system configurations in this subsection. The default values are set as follows. The transmission power is set to be 10 mW (10 dBm). The operating frequency is 10 MHz. The relay coils have the same radius of 0.15 m, and the number of turns is 20. The background noise level is −105 dBm. The permeability of the underground soil medium is a constant and is similar to the permeability of the air, since most soil in the nature does not contain magnetite. Therefore, $\mu = 4\pi \times 10^{-7}$ H/m. The soil moisture as well as other soil properties do not affect the MI communication as discussed in Section 6.2.3.

In Fig. 6.15, the channel capacities of MI waveguides with different coil resistance R and the coil interval length r are provided. Trade-offs exist here, since reducing coil resistance can reduce the path loss but also reduce the bandwidth, while the increasing the interval length between adjacent coils can increase the bandwidth but also increase the path loss. Hence, there exist optimal coil resistance and relay coil density to maximize the channel capacity, as shown in Fig. 6.15. It should be noted that although the

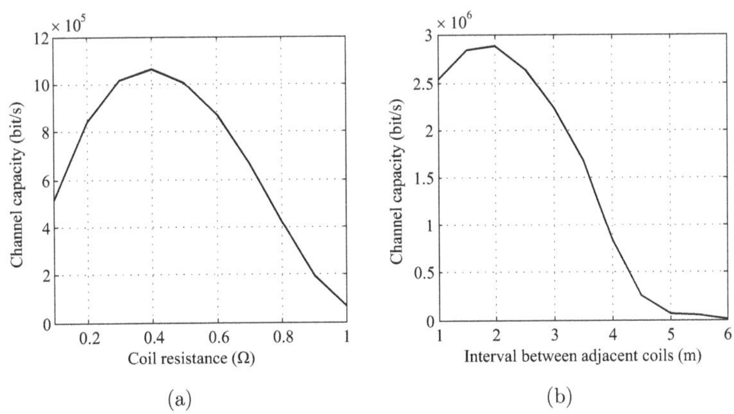

Figure 6.15 The MI channel capacity as functions of (**a**) coil resistance and (**b**) coil interval length (Note that the transmission distance is 40 m) [111]

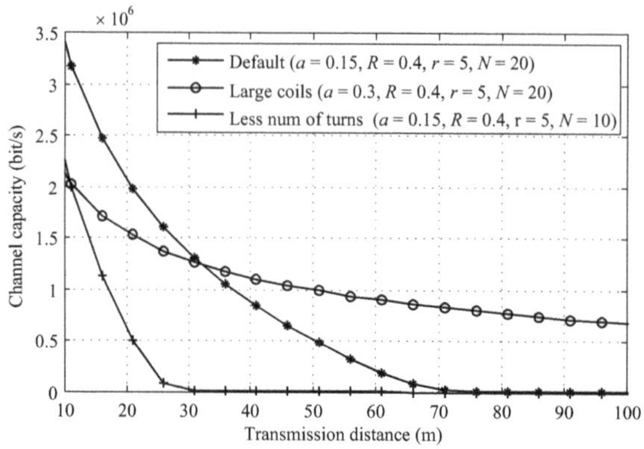

Figure 6.16 The MI channel capacity as functions of the transmission distance with different MI waveguide parameters (The coil resistance is set to the optimal value 0.4 Ω given in Fig. 6.15(a), while the coil interval is the maximum acceptable value 5 m) [111]

coil resistance can be freely set to the optimal value, the coil interval is usually set as large as possible to reduce the number of relay coils.

Figure 6.16 shows the channel capacities of MI waveguides as a function of transmission distance. The effects of different coil sizes and number of turns (as well as the operating frequency) are also captured. Different from the traditional wireless channels, both the SNR and the bandwidth dramatically decrease as the transmission distance increases. Hence, the channel capacity attenuates very fast (Mbps in 20 m vs. kbps in 70 m). This fast attenuation cannot be compensated by just increasing the transmission power since the power has no influence on the bandwidth. Increasing coil size can slightly reduce the bandwidth but can also dramatically reduce the path loss, especially in long-distance transmission. Therefore, the channel capacity with larger coil size is lower in the near region but attenuates much slower. Increasing either operating frequency or the number of turns of each coil can increase the mutual induction; hence, they have the same effects on the channel capacity.

6.3.4 Network Capacity in MI-Based WUSNs

After deriving the channel capacity between a pair of underground sensors, we investigate the achievable throughput of each node that serves as a lower bound of the network capacity in MI-based WUSNs. Since both the maximum link data rate (channel capacity) and the network geometric structure of the MI-based WUSNs is dramatically different from the traditional wireless networks, the network capacity also have significant different characteristics. The network geometric structure of the MI-based WUSNs is determined by the positions where the MI relay coils are deployed. Therefore, the network capacity is influenced by the MI waveguide deployment strategies. In [115], we proposed three types of MI waveguide deployment strategies to construct fully connected WUSNs. In this section, we analyze the network capacity of the MI-based WUSNs constructed by the three types of deployment strategies.

Network Model and Deployment Strategies

In this section, we consider a large-scale WUSN with n underground sensor nodes $\{X_1, X_2, ..., X_n\}$ randomly located in a disk of area S m^2 in the plane. Each node, equipped with an MI transceiver coil, is independently and uniformly distributed in the field. MI relay coils are deployed among the underground sensor nodes to relay the MI signal between adjacent sensor nodes, as shown in Fig. 6.14. Each underground sensor node acts as a source node and has a randomly and independently chosen destination node. Our goal is to find the mathematical expression of the achievable throughput of each node $\lambda(n)$. The achievable throughput of each node $\lambda(n)$ is defined as the maximum data rate that can be sent from each source to its destination in a multi-hop fashion under a spatial and temporal transmission scheduling scheme. Since the underground sensor nodes are a low-cost device with limit-processing ability, we assume that all transmissions employ the same power $P_t(n)$, where $P_t(n)$ can be adjusted before WUSN deployment to guarantee a certain level of communication range $d(n)$. Each node can transmit at $C(n)$ bits per second over a common MI channel. In this section, to maximize the achievable throughput, we use the channel capacity derived in the previous section as $C(n)$. We define that the minimum signal-to-interference-plus-noise ratio (SINR) for successful communication under the data rate $C(n)$ is β. Hence, transmission between the transmitter X_i and the receiver X_j is successful if and only if

$$\frac{P_t(n) \cdot 0.25\, \zeta^{-2}\left(x, \left\lceil \frac{|X_i - X_j|}{r} \right\rceil\right)}{N_0 + \sum_{k \neq i} P_t(n) \cdot 0.25\, \zeta^{-2}\left(x, \left\lceil \frac{|X_k - X_j|}{r} \right\rceil\right)} \geq \beta, \tag{6.51}$$

where x is given by (6.44), N_0 is the Gaussian noise power, and the sum in the denominator is the total interference from all simultaneously transmissions.

Then the maximum transmission rate $C(n)$ can be given by

$$C(n) = \frac{R}{\mu \pi^2 a} \sqrt{2^{\left\lceil \frac{d(n)}{r} \right\rceil^{-1}} - 1} \cdot \log(1 + \beta). \tag{6.52}$$

To facilitate the analysis, we also defined an interference range $D(n)$. Two transmitters can transmit simultaneously without interfering with each other if they are $D(n)$ apart. The relationship between the communication range $d(n)$, and the interference range $D(n)$ is determined by the MI relay coil deployment strategy and is discussed in the following subsections.

We consider three important strategies to deploy the MI relay coils to construct a fully connected WUSNs [115].

- *The spanning-tree strategy:* As shown in Fig. 6.17(a), the Voronoi tessellation of the sensor nodes (denoted as V) partitions the whole field into Voronoi cells. A spanning tree is constructed by using all sensors as vertices and all links connecting sensors in adjacent Voronoi cells in V as edges. By deploying MI relay coils along all the edges in the spanning tree, the whole WUSN is fully connected. The spanning-tree strategy uses a minimum number of relay coils but is not robust to sensor node failures.

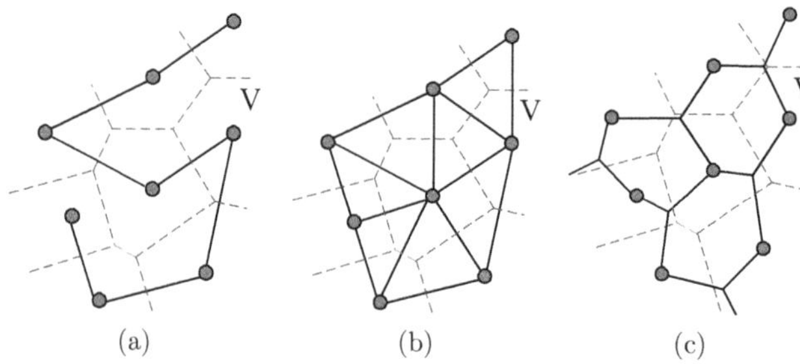

Figure 6.17 Illustration of (**a**) spanning-tree strategy, (**b**) full deployment strategy, and (**c**) triangle centroid strategy (Dots represent underground sensor nodes, solid lines represent the MI waveguide, and dashed lines are Voronoi tessellations V) [111]

- *The full deployment strategy:* As shown in Fig. 6.17(b), MI waveguides are deployed along all the edges that connect sensors in adjacent Voronoi cells in V. The full deployment strategy consumes more relay coils but is robust to sensor node failures.
- *The triangle centroid strategy:* By connecting the nodes that are in adjacent Voronoi cells in V, the whole field is partitioned into nonoverlapped triangle cells. In every other triangle cell, the MI waveguide is deployed along the three-pointed star centered at the triangle centroid, as shown in Fig. 6.17. The triangle centroid strategy achieves the same robustness to sensor node failures as the full deployment but consumes much less MI relay coils [115].

The full deployment strategy and the triangle centroid strategy can construct a WUSN with almost the same geometric structures, while the WUSN derived by the spanning-tree strategy has a completely different network geometric structure. Hence, in the following two subsections, we investigate the network capacity of MI-based WUSNs under the two network geometric structures: the structure created by the spanning-tree strategy and the structure created by the full deployment/triangle centroid strategy. It should be noted that we assume that the density of the MI relay coils remains the same, while the number of underground sensor nodes scales. In other words, as the sensor node density increases, the number of relay coils between two sensor nodes decreases since the interval between adjacent relay coils r remains same. The area S of the disk-shaped field is large enough so that the intermediate distances between the adjacent underground sensor nodes are much longer than r.

Network Capacity Scaling Law of WUSN Constructed by the Spanning-Tree Strategy

According to [116], the whole field can also be divided by a new Voronoi tessellation (denoted as V'), where each Voronoi cell contains a disk of radius $\sqrt{\frac{100S\log n}{\pi n}}$ and is contained in a disk of radius $2\sqrt{\frac{100S\log n}{\pi n}}$. It has been proved in [116] that each cell in the new Voronoi tessellation V' contains at least one underground sensor node.

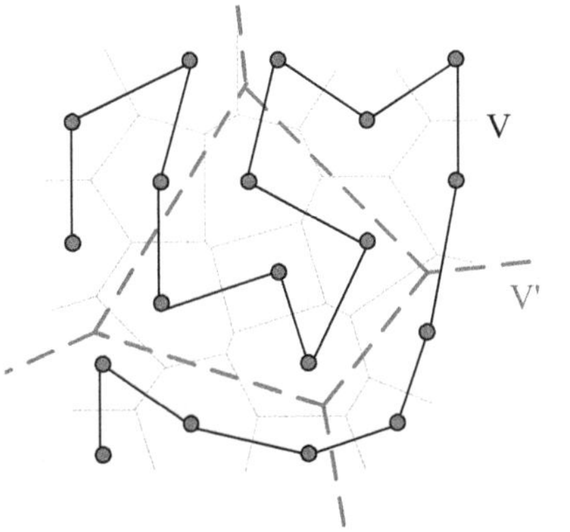

Figure 6.18 The WUSN derived by the spanning-tree strategy and two independent Voronoi tessellations V and V' constructed on such WUSN

After deriving V' (dashed lines in Fig. 6.18), the spanning-tree strategy can be applied in such WUSN to deploy MI waveguides. It is not difficult to find a spanning tree with only two leaves (i.e., a path graph denoted as G_p) connecting all underground sensors in V, while the path of the spanning tree passes each Voronoi cell in V' just once, as shown in Fig. 6.18.

The maximum distance between any points in the adjacent Voronoi cells in V' is $8\sqrt{\frac{100S\log n}{\pi n}}$. Therefore, if the transmission range of the underground sensor node is $d(n) = 8\sqrt{\frac{100S\log n}{\pi n}}$, it is guaranteed that at least one sensor node in one Voronoi cells in V' can communicate with another sensor node in the two adjacent Voronoi cells that are connected by the path graph G_p. By this means, a fully connected WUSN is derived and has a linear geometric structure. It should be noted that a sensor node in a Voronoi cell in V' can only communicate with (or interfere) another Voronoi cell through the path graph G_p where the MI relay coils are deployed.

Lemma 6.1 *The interference range $D(n)$ under the spanning-tree deployment strategy is*

$$D(n) = r\log_x(2\beta) + 2d(n), \tag{6.53}$$

where $d(n) = 8\sqrt{\frac{100S\log n}{\pi n}}$ is the communication range, r is the fixed interval length between two adjacent relay coils, and x is given in (6.44)

Proof Since the interference range is $D(n)$, the disks of radius $0.5D(n)$ around each simultaneously transmitting node are disjoint. In other words, on the path graph G_p, every simultaneously transmitting node is $D(n)$ apart from each other. Then the SINR at a receiver X_j is lower bounded by

$$\frac{P_t(n)\cdot 0.25\cdot \zeta^{-2}\left(x,\left\lceil \frac{d(n)}{r}\right\rceil\right)}{N_0+\sum_{k=1}^{+\infty}2\cdot P_t(n)\cdot 0.25\cdot \zeta^{-2}\left(x,\left\lceil \frac{0.5k\cdot D(n)}{r}\right\rceil\right)}
= \frac{\frac{P_t(n)}{N_0}}{4\zeta^2\left(x,\left\lceil \frac{d(n)}{r}\right\rceil\right)+2\frac{P_t(n)}{N_0}\cdot\sum_{k=1}^{+\infty}\left[\frac{\zeta\left(x,\left\lceil \frac{d(n)}{r}\right\rceil\right)}{\zeta\left(x,\left\lceil \frac{0.5k\cdot D(n)}{r}\right\rceil\right)}\right]^2}. \tag{6.54}$$

Note that the coefficient 2 in the denominator is added since there may exist two simultaneous transmitters on each side of the current transmitter on the path graph G_p. In sum in the denominator in (6.54), each addend is a fraction. According to the definition of $\zeta(x,k)$ given in (6.40), the denominator $\zeta\left(x,\left\lceil \frac{0.5k\cdot D(n)}{r}\right\rceil\right)$ is, in fact, a weighted sum of the numerator $\zeta\left(x,\left\lceil \frac{d(n)}{r}\right\rceil\right)$, where the weight consists of $\left\{1,x,x^2,\ldots,x^{\left\lceil \frac{0.5k\cdot D(n)}{r}\right\rceil-\left\lceil \frac{d(n)}{r}\right\rceil}\right\}$. Similar to the analysis in Section 6.3.3, since x is relatively large, the highest-order variable of x has the most influence. Therefore, the fraction in the sum in (6.54) can be upper bounded by

$$\frac{\zeta\left(x,\left\lceil \frac{d(n)}{r}\right\rceil\right)}{\zeta\left(x,\left\lceil \frac{0.5k\cdot D(n)}{r}\right\rceil\right)} < x^{-\left(\left\lceil \frac{0.5k\cdot D(n)}{r}\right\rceil-\left\lceil \frac{d(n)}{r}\right\rceil\right)}. \tag{6.55}$$

Since the transmission power $P_t(n) \gg N_0$, the SINR at a receiver X_j is approximately lower bounded by the following value that should be larger than the threshold β:

$$\frac{1}{2\sum_{k=1}^{+\infty}x^{-2\left(\left\lceil \frac{0.5k\cdot D(n)}{r}\right\rceil-\left\lceil \frac{d(n)}{r}\right\rceil\right)}} \geq \beta. \tag{6.56}$$

Therefore, the safe interference range $D(n)$ to derive the required SINR is

$$D(n)=\frac{r}{\log x}\log\left(1+2\beta\cdot x^{2\left\lceil \frac{d(n)}{r}\right\rceil}\right)\simeq r\log_x(2\beta)+2d(n). \tag{6.57}$$

□

A Voronoi cell in V' is an interfering neighbor of another cell if the minimum distance between the two cells along the path graph G_p is smaller than the interference range $D(n)$.

Lemma 6.2 *Every cell in* V' *has no more than* $44+\frac{16r\log_x(2\beta)}{d(n)}$ *interfering neighbor cells.*

Proof The Voronoi cells can only interfere with each other through the MI waveguides on the path graph G_p. Hence, it only needs to calculate how many Voronoi cells can be accommodated along the path graph G_p within the interference range. All the Voronoi cells in V' are contained by a disk of radius $\frac{1}{4}d(n)$, and the interference range is $D(n)$. All interfering neighbor cells have to be within a line segment of length $2*[3*\frac{1}{4}d(n)+D(n)]$. Additionally, since each cell in V' contains a disk of radius $\frac{1}{8}d(n)$, the line segment cannot cross more than $\frac{2*[3*\frac{1}{4}d(n)+D(n)]}{\frac{1}{8}d(n)}$ cells in V', which completes the proof.

□

After deriving the upper bound of the number of interfering neighbors, a temporal transmission schedule can be designed to guarantee the success of each transmission. Specifically, each duty cycle is divided into $45+\frac{16r\log_x(2\beta)}{d(n)}$ slots, and each Voronoi cell in V' gets one slot to transmit data. Since all the Voronoi cells are along the path graph G_p, we just need to periodically allocate the $45+\frac{16r\log_x(2\beta)}{d(n)}$ slots to the Voronoi cells in V' along the path graph G_p in serial order. Since the maximum number of interfering neighbor cell is $45+\frac{16r\log_x(2\beta)}{d(n)}$, the sensor node in every cell can successfully receive the data from a transmitter within the distance $d(n)$ by using the above transmission schedule.

Since each Voronoi cell in V' contains at least one underground sensor node and the sensor nodes in each cell are guaranteed to be able to communicate with the adjacent sensor nodes along the path graph G_p, for any source–destination pair, there exists a route along the path graph G_p that connects the source node and the destination node.

Before deriving the network capacity, the traffic load of each Voronoi cell in V' needs to be investigated first. We consider the route R_{sd} from the source X_s to the destination X_d and check.

Lemma 6.3 *The probability that the route R_{sd} intersects any one Voronoi cell v in V' is upper bounded by*

$$P(R_{sd}\ intersects\ v) \leq \frac{1}{2} + \frac{c_0 \log n}{n}, \tag{6.58}$$

where c_0 is a constant.

Proof As discussed previously, all underground sensor nodes are along a path graph G_p. We can sequentially denote those node as $\{X_i, i = 1, \ldots, n\}$ along the path. The source node X_s and the destination node X_d are among the ordered sensor nodes. Assuming that the Voronoi cell v contains the mth sensor nodes X_m. Due to the linear network structure, R_{sd} intersects v if and only if X_m is in the middle of X_s and X_d on the path graph. Since the source X_s and the destination X_d are uniformly selected from $\{X_i, i = 1, \ldots, n\}$,

$$\begin{aligned} &P(R_{sd} \text{ intersects } X_m) \\ &= P(s < m) \cdot P(d > m) + P(s > m) \cdot P(d < m) \\ &= 2\frac{m-1}{n}\frac{n-m}{n}. \end{aligned} \tag{6.59}$$

$P(R_{sd}$ intersects $X_m)$ in (6.59) achieves the maximum value if $m = \lceil \frac{n}{2} \rceil$. Note that one Voronoi cell in V' may contain multiple sensor nodes. The number of nodes in each cell is lower bounded by $c_l \log n$ and upper bounded by $c_u \log n$, according to [116]. Then,

$$\begin{aligned} &P(R_{sd} \text{ intersects } v) \\ &< \max\{P(R_{sd} \text{ intersects } X_m)\} + 2P(X_i \text{ is inside } v) \\ &\leq \frac{1}{2} + 2\frac{c_u \log n}{n}. \end{aligned} \tag{6.60}$$

This completes the proof. □

Since the number of total routes is n (every node acts as source and one route for one source-destination pair), the mean number of the routes served by each cell is

$$E[\text{Number of routes intersecting } v] < \frac{n}{2} + c_0 \log n. \tag{6.61}$$

Next, we use the Vapnik–Chervonenkis theorem, a similar strategy to the one used in [116], to prove that the actual number of routes intersecting a cell converges to the above derived mean value. The proof procedure is similar to the method in [116]; hence, it is omitted here. Then, we have

Lemma 6.4 *There exists a $\delta(n) \to 0$ as $n \to \infty$ such that*

$$P\left[\sup_{v \in V'}(\text{Number of routes intersecting } v) \leq \frac{n}{2} + c_0 \log n\right] \geq 1 - \delta(n). \tag{6.62}$$

By using the transmission schedule scheme given previously, each Voronoi cell in V' can successfully transmit data in the allocated time slot. Since the channel capacity is $C(n)$ given by (6.52), the data rate each cell can use is the channel capacity $C(n)$ divided by the total number of time slots $45 + \frac{16r \log_x(2\beta)}{d(n)}$. Meanwhile, the traffic load for each Voronoi cell in V' is given by the throughput of each node $\lambda(n)$ multiplied by the actual number of routes served by each cell. In the most efficient case, the traffic load per cell is equal to the available data rate. Hence, we have

$$\lambda(n) \cdot \left(\frac{n}{2} + c_0 \log n\right) = \frac{C(n)}{45 + \frac{16r \log_x(2\beta)}{d(n)}}. \tag{6.63}$$

Then the achievable throughput of each underground sensor node under the spanning-tree strategy in MI-based WUSNs is derived.

Theorem 6.5 *For MI-based WUSN constructed by the spanning-tree strategy, the achievable throughput of each underground sensor node is*

$$\begin{aligned}\lambda(n) &= \frac{2}{n + 2c_0 \log n} \cdot \frac{R}{\mu\pi^2 a}\sqrt{2^{\lceil \frac{d(n)}{r} \rceil^{-1}} - 1} \cdot \frac{\log(1+\beta)}{45 + \frac{16r \log_x(2\beta)}{d(n)}} \\ &\simeq c_1 \cdot \frac{1}{n + 2c_0 \log n} \cdot \left(2^{c_2 \sqrt{n/\log n}} - 1\right)^{1/2},\end{aligned} \tag{6.64}$$

where c_1 and c_2 depend only on the SINR threshold β and the MI waveguide parameters but do not depend on the sensor node number n.

Compared with the achievable throughput in traditional wireless networks $(\frac{c}{\sqrt{n \log n}})$ [116], we can find dramatical differences in the throughput of MI-based WUSNs in (6.64).

Network Capacity Scaling Law of WUSN Constructed by the Full Deployment Strategy and Triangle Centroid Strategy

The WUSN constructed by the full deployment strategy or the triangle centroid strategy has completely different network geometric structure than that of the spanning-tree strategy. As a result, the achievable network capacities of the former two deployment strategies are also different.

As shown in Figs. 6.17(b) and (c), in the full deployment strategy and the triangle centroid strategy, the MI relay coils are deployed to connect all the underground sensors in adjacent Voronoi cells in V. The only different is the route of the MI waveguide: The triangle centroid strategy places the MI waveguide along the three-pointed star in every other cell, while the full deployment strategy places the relay coils along all the edges of the triangle cells. However, the network connectivity and topology are exactly the same for these two deployment strategies. Hence, they share the same network capacity scaling law.

Since each underground sensor nodes can directly communicate with all its adjacent sensor nodes, the communication and interference range of such sensor nodes are almost isotropic, which yields the network geometric structure just like the traditional wireless networks. Therefore, the results derived in [116] can be unitized but need modifications. Due to the usage of the MI waveguide, the channel capacity and the interference model are different from the traditional wireless networks. Therefore, the network capacity of such WUSNs can be derived by using similar strategies to that used in Section 6.3.4 and the results in [116].

Theorem 6.6 *For the MI-based WUSN constructed by the full deployment strategy or the triangle centroid strategy, the achievable throughput of each underground sensor node is*

$$\lambda(n) \simeq \frac{c_3}{\sqrt{n \log n}} \cdot \left(2^{\,c_2 \sqrt{n/\log n}} - 1\right)^{1/2}, \tag{6.65}$$

where c_2 and c_3 are the variables depending only on the SINR threshold β and the MI waveguide parameters.

Comparing (6.65) with the achievable throughput in traditional wireless networks and in WUSNs constructed by the spanning-tree strategy, we can find that the throughput of MI-based WUSNs in (6.65) attenuates significantly slower than that in (6.64) and the $(\frac{c}{\sqrt{n \log n}})$ in [116].

6.3.5 Numerical Analysis

In Fig. 6.19, the network capacity scaling laws of the traditional ad hoc networks and the MI-based WUSNs are compared. For fairness, the normalized achievable throughput $\overline{\lambda(n)} = \lambda(n)/\lambda(n_0)$ is plotted, which can eliminate the effects of constants that do not depend on the node number n. It should be noted that the scaling laws of the MI-based WUSNs is a function of the variable $c_2 \sqrt{n/\log n}$ in (6.64) and (6.65), which is determined by $d(n)$ and r. The constant c_2 needs to be selected before the numerical comparison. Since we assume that intermediate distances between adjacent underground sensor nodes should be longer than the coil interval r, c_2 can be selected accordingly. In Fig. 6.19, we let $c_2 = \sqrt{\frac{\log 1000}{1000}}$. Figure 6.19 shows that the network

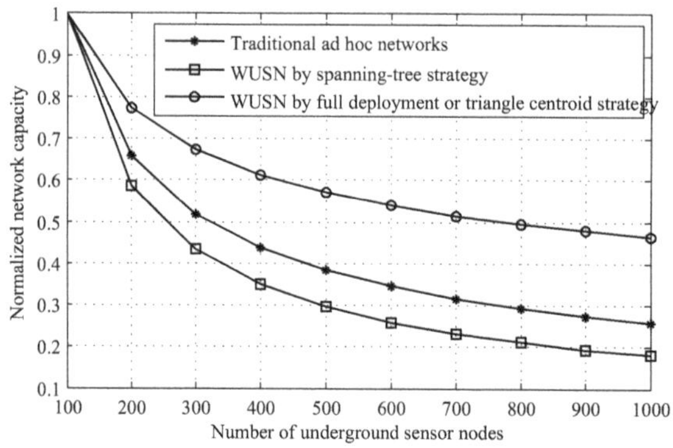

Figure 6.19 The network capacity scaling laws of the traditional ad hoc networks, the MI-based WUSNs constructed by the spanning-tree strategy, and the MI-based WUSNs constructed by the full deployment or triangle centroid strategy [111]

capacity of WUSNs constructed by the spanning-tree strategy has a similar decreasing speed to the traditional ad hoc network as the node number increases, which is the joint effect of two factors: The $\frac{1}{n+2c_0 \log n}$ in (6.64) drops much faster than the $\frac{1}{\sqrt{n \log n}}$ of traditional networks; however, the $(2^{c_2 \sqrt{n/\log n}} - 1)^{1/2}$ in (6.64) can dramatically compensate this fast attenuation. Figure 6.19 also indicates that the network capacity of WUSNs constructed by the full deployment or the triangle centroid strategy falls much slower than the traditional ad hoc networks as the node number increases due to the following reason. Although these two types of networks have almost the same geometric structure, the network capacity of the MI-based WUSNs can benefit from the increased node density since the MI channel bandwidth increases dramatically as the node intermediate distance decreases.

6.3.6 Discussion on the Reliability of the Channel and Network Capacities in MI-Based WUSNs

In this section, we investigate the impact of several factors on the channel and network capacities in MI-based WUSNs derived in the previous two sections. The factors that we studied include the sensor node failure, relay coil missing, and direction/position deviation of the relay coils. The influence of these factors on the channel and network capacities reflects the system reliability of the MI-based WUSNs under different deployment strategies.

Unreliability Due to Node Failure

In harsh underground environments, sensor node failure may frequently happen due to device damage, malfunction, or dead battery. The sensor node failure cannot affect the channel capacity of the remaining links. However, node failure can change the network geometric structure. Hence, the network capacity can be influenced by the node failure.

Network capacities of WUSNs constructed by different MI waveguide deployment strategies have different robustness to node failure.

- In WUSNs constructed by the spanning-tree strategy, the failure of any one node will partition the whole network into two disconnected networks. According to (6.61), by deleting one node, up to half of the routes are disconnected. As a result, up to 50% of traffic load cannot be delivered since there is only one route connecting each source–destination pair. Consequently, if only one sensor node fails, the achievable network capacity can drop to as low as 50% of the original value. Therefore, the node failure can significantly affect the reliability of the network capacity of the MI-based WUSNs constructed by the spanning-tree strategy.
- In WUSNs constructed by the full deployment strategy or triangle centroid strategy, each sensor node is connected to all its adjacent neighbors. The failure of a few sensor nodes cannot partition the network into unconnected parts. If the remaining source and destination nodes are aware of the failure of any node in the network, it is not difficult to find other near routes to avoid the failed sensor nodes. This may slightly rise the traffic load of nearby sensor nodes. However, consider the large number of sensor nodes in the investigated large-scale network, the change of the average achievable throughput of each node can be neglected. Therefore, the node failure has little influence on the reliability of the network capacity of the MI-based WUSNs constructed by the full deployment strategy or triangle centroid strategy.

Unreliability Due to Relay Coil Missing and Direction/Position Deviation

As discussed previously, the MI waveguide consists of multiple MI relay coils. The channel model in Section 6.3.3 assumes that all relay coils are placed at the designed position at right angle. However, in a practical situation, this assumption may not hold since: (1) the relay coils may be damaged by the harsh underground environments and (2) the position and direction of each relay coils may deviate due to the environmental perturbations.

The relay coil missing and direction/position deviation cannot disconnect a wireless link immediately. Hence, those factors cannot change the network geometric structure. However, the unreliability caused by the relay coils can gradually reduce the channel capacity and consequently affect the network capacity. In the following, we reexam the channel capacity under the influence of the relay coil missing, the position deviation, and the direction deviation.

Due to the above three factors, the mutual induction defined in (6.41) between each pair of adjacent relay coils $\{M_i, i = 1,2,...,n\}$ is no longer the same. In particular, the missing of a certain relay coil can cause a certain coil interval doubled and the total relay number $\lceil \frac{d}{r} \rceil$ decreased by 1. The coil position deviation will cause the fluctuations of the coil interval r along the MI waveguide. The coil direction deviation may cause the θ_t and θ_r become a value other than $0°$. Moreover, since the mutual inductions $\{M_i, i = 1,2,...,n\}$ are no longer the same, the path loss in (6.39) needs to be modified as

$$L'_p(k,f) \simeq 0.25 \cdot \left[(2\pi f)^k \cdot \prod_{i=1}^{k} M_i\right]^2 \cdot |\zeta'_k|^{-2}, \tag{6.66}$$

where $k - 1$ is the number of functional relay coils between the two underground sensors; the polynomial ζ'_k is defined as

$$\begin{aligned}\zeta'_1 &= Z,\\ \zeta'_2 &= Z^2 + 4\pi^2 f_0^2 M_1^2,\\ &\vdots\\ \zeta'_k &= Z \cdot \zeta'_{k-1} + 4\pi^2 f_0^2 M_{k-1}^2 \cdot \zeta'_{k-2}.\end{aligned} \tag{6.67}$$

Once the new path loss is derived, the new 3-dB bandwidth B' can be calculated by letting the path loss at the frequency $f_0 + 0.5B$ be the twice as much as the path loss at the central frequency f_0, i.e., $L'_p(k,f_0 + 0.5B') = 0.5L'_p(k,f_0)$.

Due to the randomness of the coil missing and the direction/position deviation, it is impossible to derive the closed-form expression of the channel capacity under those factors. In the following, we conduct numerical evaluations on the impact of relay coil missing and direction/position deviation. It should be noted that the impact of those factors on the channel capacity is proportional to their impact on the network capacity, since neither the coil missing nor the direction/position deviation can change the network's geometric structure. Therefore, we focus on the impact of those factors on the channel capacity.

Using the same default parameters given in Section 6.3.3, we consider three scenarios: (1) each relay coil has a probability of 10% to get damaged and inability to function; (2) the direction of each relay coil θ deviates from 0° by x_d, where x_d is the zero mean gaussian variable with a standard deviation of $10\% \times 90°$; and (3) the position of each relay coil deviates from the original position by x_p, where x_p is the zero mean gaussian variable with a standard deviation $10\% \times r$. The results shown in Fig. 6.20 are the average of 100 iterations. Figure 6.20 shows that the coil

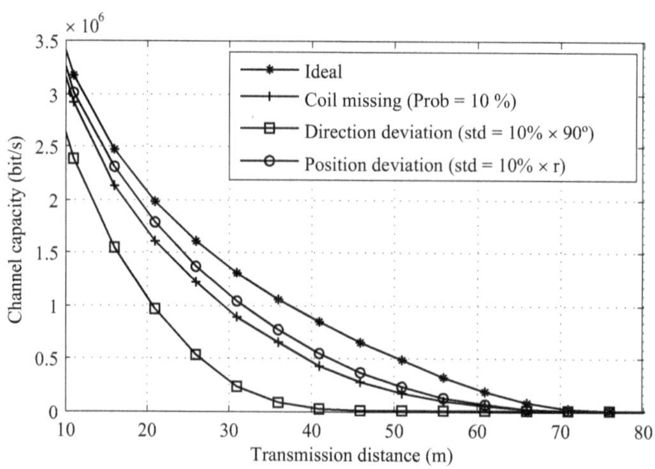

Figure 6.20 The channel capacity of MI-based WUSNs under the impacts of coil missing, direction deviation, and position deviation [111]

direction deviation has the most severe impact on the channel capacity (as well as the network capacity) of the MI-based WUSNs since the consecutive magnetic inductions depend on the proper direction of all relay coils. Meanwhile the impact of the position deviation is not significant since it only causes mild fluctuations of mutual inductions.

7 Magnetic Communication Networking Protocol Stacks

Magnetic communication is a novel physical communication technology. To network a large number of MC devices, traditional networking protocols can be employed, but we need to make significant modifications on the physical layer (PHY) to accommodate the special features of MC. For example, to access the communication medium, traditional carrier sensing multiple access (CSMA) or ZigBee can be adopted, but MC has a short communication range, and antennas of different devices may have substantial coupling that can affect the communication performance. To address this issue, we need dedicated scheduling algorithms to reduce mutual couplings among coils and increase the network throughput.

In this chapter, we first introduce a complete MC network stack. Then, we show unique features of MC at the network level. After that, we introduce the scheduling algorithms for MC networks.

7.1 Overview of Magnetic Communication Networking Protocol Stacks

Since MC is a fundamental wireless communication mechanism, the networking protocol stacks depend on the specific application as well as the manner wireless devices are connected together. For example, sensors that are buried in underground soil can communicate at a very low data rate, and the required communication range for them is long. In such a case, the network has to support the long range and low data rate. When MC is used for smart payment, the communication range has to be very short and the network must only consist of the sender and the destination. Thus, depending on the specific application, we can use MC in different ways.

Although a network may demonstrate different structures, networking protocols follow a general architecture that is the same as existing wireless networking protocols. Next, we use NFC as an example to introduce networking protocol stacks.

Overall, networking protocol stacks consist of a PHY, a data link layer, a network layer, a transport layer, and an application layer. Although according to the Open Systems Interconnection (OSI) networking model [121], there are two more layers, namely, the session layer and the presentation layer that are not widely used here.

7.1.1 Network Protocol Stacks

First, the PHY is used to build connections between wireless devices. When there are data being sent to a destination, the PHY can convert binary 0s and 1s to wireless waveforms by using coding and modulation, and then send the waveforms to the antenna that can radiate the signals into the space. Once the receiver receives the wireless signals, it can perform bit synchronization, frequency synchronization, demodulation, and decoding. Then, the PHY sends the decoded 0s and 1s to the data link layer. In previous chapters, we have discussed MC modulation and antenna design that can be used for this PHY. Also, the physical layer can involve bit rate control and transmission mode control. For example, the wireless radio can adjust the communication bit rate by observing the wireless channel. When the channel is good, it can transmit signals using a high data rate, while when the channel is poor, it can transmit signals using a low data rate. In this way, the bit error rate can be reduced, and the wireless communication becomes very reliable. Recently, the development of full-duplex radios opens another door for wireless communications. Thus, besides traditional half-duplex radio, the MC can also adopt full-duplex radios provided that the hardware can support this new technology.

The data link layer aims to reduce the transmission errors between two wireless devices. It provides framing services to incoming data. Framing can include special bits at the beginning or end of the frame. Also, the data link layer adds sender's and receiver's physical addresses to the frame. To ensure reliability, the data link layer uses error control schemes that can detect errors and request for retransmissions. An important task of the data link layer is the medium access control. Due to the broadcasting nature of the wireless communication, interference can be significant when multiple wireless devices transmit. Therefore, we need to control their access to the wireless channel. The data link layer is responsible for this, and it controls the access to the medium by using different schemes such as CSMA, time division, and frequency division, among others. Another issue that the data link layer can solve is the flow control. The data rate sent out from the sender should be the same as the data rate at the receiver in order to avoid errors. The data link layer can coordinate with the sender and the receiver to follow the constant data rate.

The network layer is responsible for identifying each wireless device by using a unique IP addresses and find an optimal route to send the data packet from the sender to the destination. Generally, its function includes routing and logical addressing. In a large wireless network, routing is a critical problem, which can introduce additional delay, energy consumption, and traffic in the network if the algorithm is not well designed.

The transport layer serves the application layer. It receives data from the application layer, and then it segments the data and adds more detailed information about how to send the data packet. When it receives data from the network layers, it extracts useful information and passes it to the application layer.

The application layer is dedicated for specific applications. Users can use the applications without knowing what is going on in the transport layer, network layer, data link layer, and PHY. The application layer generates data, which initiates the data transmission.

The MC networking protocol stacks can follow this architecture. Existing networking protocols such as Zigbee can also be used to build networks for MC.

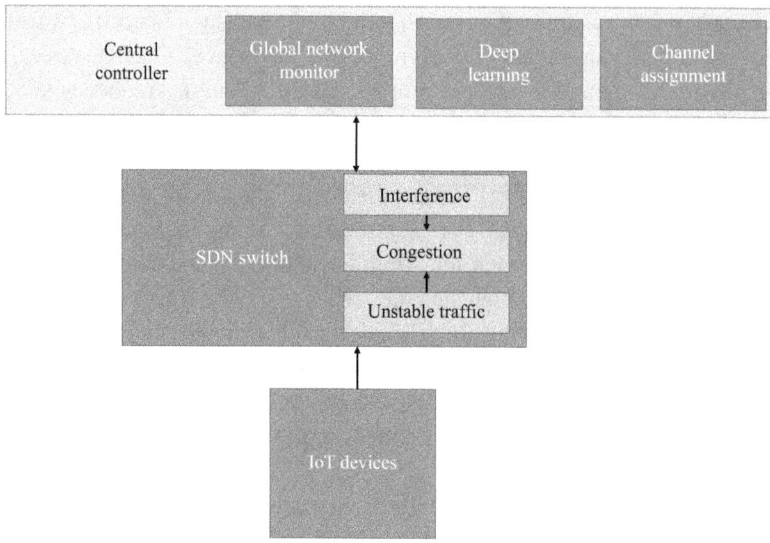

Figure 7.1 A simplified architecture of SDN-IoT

7.1.2 Software-Defined Networking for Magnetic Communication Networks

Different from wireless cellular communications or wireless local area networks, which serve a large population and require very high data rate, MC is used for secure applications and IoT in extreme environments. Usually, there are no standalone networks that are dedicated for MC. As a technology for IoT, the generated data are routed to the existing networks in the end. The Open Systems Interconnection (OSI) network model introduced in the preceding section can support a single physical communication mechanism well. However, when a wireless device is equipped with multiple wireless communication mechanisms and the data come from various sources, the OSI network model may experience various challenges in integrate and frame data packets.

One of the promising solutions to integrate data from networks using MC is the recently developed software-defined networking (SDN) [122], which has been extensively used for IoT [123, 124]. Figure 7.1 shows a simplified architecture of the SDN-IoT, where the network is divided into three planes, namely the sensing plane, data plane, and control plane. The detailed architecture can be found in [123]. In the sensing plane, sensors using MC can sense the information of interest and transmit data to the network switches. These IoT devices can be very simple; they do not have a complete networking stack. They can be equipped with sensors, microcontrollers, and communication units. In the data plane, the network switches can be connected using wires or wireless technologies. The data traffic, interference, and congestion can be managed. The control plane takes care of resource allocation.

The SDN-IoT has a significant advantage over traditional networking protocols since it can aggregate data from various sources with different communication protocols and ensure that the network being reliable, robust, and efficient. MC may serve a small amount of wireless devices, and it does not have a standalone network. Therefore, the SDN-IoT technique can accommodate wireless devices using MC.

7.1.3 MI Communication Protocol Design: An Example

An example of the MI communication and networking protocol is shown in [125]. The communication and networking protocols are designed mainly by using the software-defined radios. MI antennas are enhanced by using metamaterials and are directly connected to universal software radio peripheral (USRP) boards. The mother board is the USRP N210, which is based on a Xilinx Spartan-3A DSP 3400 FPGA, and the daughter boards are LFTX/LFRX, which can support two independent antennas through connectors TxA/RxA and TxB/RxB. The daughter board can generate and receive wireless signals from 0 Hz to 30 MHz. Such a frequency band can fully support the development of MI communication systems. By using the aforementioned equipment, the signal is generated and analyzed in GNU radios.

This protocol is mainly used to support MI-based wireless sensor networks in extreme environments, such as the underground environment. In the PHY, since the MI communication has a narrow bandwidth, the continuous-phase modulation (CPM) is used to minimize the out of band power leakage. The CPM has a limited out of band power, which makes it spectral efficient. For the symbol synchronization, the maximum-likelihood algorithm is used to detect the exact starting point of the symbol. Moreover, channel estimation and equalization are performed to optimally decode the signal. The channel is first estimated using a training-based approach and then conduct linear adaptive equalization using the zero-forcing algorithm. The BCH codes are also utilized to do forward error correction to reduce errors. In the data link layer, a master–slave network configuration is used, and CSMA is used for medium access control. In the network layer, an adaptation layer is defined to integrate the IPv4 protocol. Based on this, an MI communication and a networking protocol are developed.

7.2 Point-to-Point Network Protocol

In the previous section, we introduce the fundamental network protocols for MCs. Also, we provide examples of MC and networking protocol design. Since MC has a very limited communication range, it is challenging to form a large-scale network by solely using MC. As a result, heterogeneous networks are promising solutions that can combine the advantages of MC's low power consumption and reliable channel with legacy networks' large coverage and high throughput. In this section, we introduce the short-range point-to-point MC. Particularly, we will focus on the existing NFC and use it as an example to show the unique characteristics of MC networks. In the next section, we will focus on multiple transceivers using only MC and the communication range extended by using the magnetic waveguide.

7.2.1 PHY Characteristics of Magnetic Communication

Based on previous discussions, we have learned that the physical layer (PHY) of MC has the following advantages and disadvantages. First, its wireless channel is reliable because it uses low carrier frequency, which results in a long wavelength. Thus, for the NFC with a range smaller than 0.1 m, it is barely affected by the multipath fading

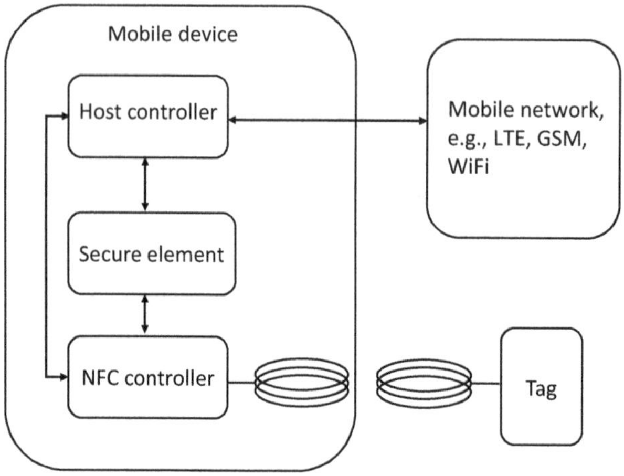

Figure 7.2 An illustration of the network architecture of NFC modules

or shadowing. Second, the power consumption of MC is extremely low compared with existing wireless solutions, such as WiFi and Bluetooth. Since the MC is based on wireless communication operating at relatively low frequency, the same technology can be used for energy harvesting. The device can operate at a standby mode by turning off its power supply, and thus it does not consume any power. Once it is enquired by a reader, it can harvest energy first to turn on its power supply to move into the active mode. This is fundamentally different from other short-range communication protocols that have nonzero standby power consumption. Third, MC operates at frequency bands at/lower than the HF band. Modern-day electronics can support the fabrication of such radios with extremely low cost. For instance, by using simple microcontrollers, we can perform all the signal processing for MC [126]. Most of the high-cost components that are used for microwave communications can be saved.

However, the MC has the following disadvantages. First, its communication range is short. Compared with WiFi, Bluetooth, Zigbee, and RFID, the communication range of MC is extremely short. Therefore, the coverage of MC is poor. Using MC to build a network for connecting smart devices such as sensors is challenging. It requires a highly dense network with a large number of transceivers with MC capability, which is not practical. For example, the NFC has a short communication range. We cannot use it to form a large-scale network. As shown in Fig. 7.2, the NFC module is used for secure applications in the vicinity of the user, and the data is transmitted to Internet by using cellular networks. As a result, the point-to-point MC can only be used within a short range, which can be considered as an advantage, i.e., secure, as well as a disadvantage, i.e., a short range.

7.2.2 Network Stack of Near-Field Communication

NFC is the most widely used point-to-point MC. Thus, in this section, we will mainly introduce its networking stacks and protocols. NFC is intentionally designed with a short communication range to enable the touch-then-communicate capability.

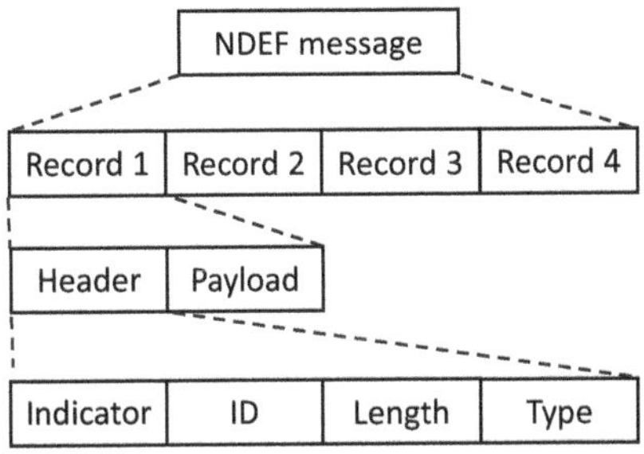

Figure 7.3 An illustration of the NDEF message

It has four different operation modes that depend on the actual use case; they are reader/writer mode, peer-to-peer mode, card emulation mode, and wireless charging mode. The wireless charging mode is relatively new compared with the other three modes. It allows a device to be charged using the same coil that is used for NFC. Also, NFC can be used to coordinate the wireless charging. For example, by using NFC, a device can communicate with a charger to update its battery information. The reader/writer mode allows the reader to read or write into a tag. This mode is similar to the RFID (radio frequency identification). The peer-to-peer mode is used for the communication between two NFC devices. For example, it can be used to exchange the WiFi authentication information between a device and a WiFi access point instead of searching for the WiFi access point and manually typing the password. The card emulation mode is used to replace a physical card/ticket. It can be used as a credit card with secure information saved in the NFC device.

The data communication in NFC relies on the NFC Forum Data Exchange Format (NDEF), which is in a lightweight binary message format. It is designed to encapsulate multiple payloads that are defined by specific applications into one message. The NDEF message consists of one or multiple NDEF records, which can carry payloads of different types. A large payload can be divided into multiple small payloads. There are three key parameters that are used to define a payload, namely the payload length, the payload type, and the payload identifier. An illustration of the NDEF message structure is shown in Fig. 7.3.

The NDEF record layout is shown in Fig. 7.4. The MB bit stands for message begin, which is a 1-bit field indicating the start of an NDEF message. The ME bit stands for message end, which is a 1-bit field indicating the end of an NDEF message. The CF bit stands for chunk flag, which is a 1-bit field indicating that the record is either the first or a middle one. The SR bit stands for short record, which is a 1-bit field indicating the payload length filed is a single octet. The IL bit stands for the ID-LENGTH field, which indicates whether the ID-LENGTH field is omitted or not. The TNF stands for the type name format. After setting these headers, the information about the payload, such as length, ID, and type, can be specified. The payload is at the end of a record. More detailed information can be found in [127].

NFC has different protocol stacks for different operation modes. Here, we provide an example of the reader/writer mode protocol stack, an illustration of which is shown

7	6	5	4	3	2	1	0
MB	ME	CF	SR	IL	TNF		
TYPE LENGTH							
PAYLOAD LENGTH 3							
PAYLOAD LENGTH 2							
PAYLOAD LENGTH 1							
PAYLOAD LENGTH 0							
ID LENGTH							
TYPE							
ID							
PAYLOAD							

Figure 7.4 NDEF record layout [127]

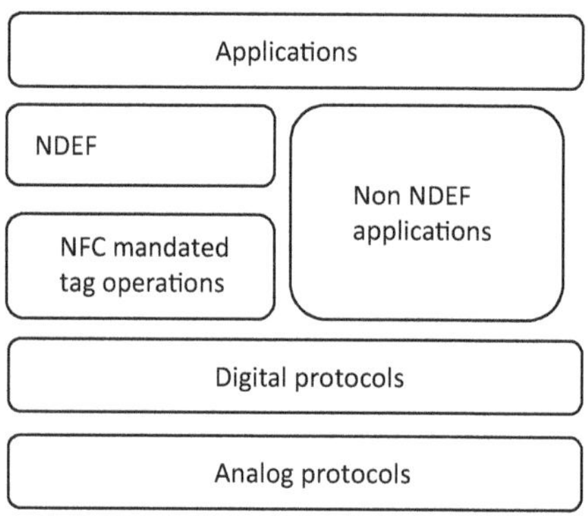

Figure 7.5 NFC protocol stack for the reader/writer mode

in Fig. 7.5. The analog and the digital protocols are the PHY technologies. The analog protocols include the RF components of NFC devices that affect the communication performance. Digital protocols include the digital components of the communication system, such as collision detection. Tag operations include the commands and instructions for NFC Forum-mandated tags, namely, from the Type 1 to Type 4. Applications are based on the NDEF and non-NDEF formats, which depends on vendors' specifications.

7.3 Point-to-Multipoint Network Protocol

In the previous section, we present the point-to-point MC for NFC. Due to the security issue, NFC has very limited communication range, and only point-to-point communication is allowed. Another widely used MC is RFID. Two of its operating frequency bands use MI instead of propagating electromagnetic waves, which are the 13.56 MHz band and 125–135 kHz band. An RFID reader can read passive tags within a short range. Different from the NFC, RFID's communication range is longer, i.e., from 0.1 to 2 m. The UHF and microwave RFID can read passive tags in a longer range, as far as 7 m. However, the communication mechanism is fundamentally different and, thus, in the section, we only focus on the short-range RFID at 13.56 MHz and lower than 135 kHz bands.

RFID has a reader that initiates the communication with tags [128, 129]. Tags have identification information that can be used to identify items. Also, tags can be integrated with sensors to provide both identification and sensing functions [126, 130]. Wireless information is modulated by changing tag's impedance and backscatter the wireless continuous wave that is generated by the reader. Note that, since the RFID using MI does not generate electromagnetic waves, some literature studies only consider the backscatter as a concept of electromagnetic wave, and it is solely induction for RFID at 13.56 MHz and below 135 kHz. Here, we consider in the near field that the impedance modulation also creates backscattered signals since the reader is affected by the impedance modulation.

The 13.56 MHz RFID has a communication range of 10–20 cm. Its coil is relatively small with no more than 10 turns. It is used for access control, smart cards, and libraries. The LF band RFID with carrier frequency lower than 135 kHz has a range of 2 m. It has a large coil with more than 100 turns. Its data rate is slow, which is lower than 10 kbps. It can be used for animal tagging and vehicle identification.

The RFID network stack mainly has two layers, i.e., the physical layer and the tag-identification layer. The PHY includes the antennas and RF signal processing. The amplitude shift keying (ASK) is used as the modulation scheme, including the double sideband amplitude shift keying, single-sideband amplitude shift keying, and phase-reversal amplitude shift keying using a pulse-interval encoding format. For the downlink communication from the reader to the tag, the reader sends information to one or multiple tags by modulating RF signals. For the uplink communication from the tag to the reader, the reader sends out unmodulated continuous wave and receive backscattered signals from the tag. Tags may employ amplitude- or phase-based modulation schemes. The tag uses FM0 or a Miller-modulated subcarrier as the encoding format to respond to the reader's command.

The tag-identification layer defines the way to manage tag populations by a reader, which includes select, inventory, and access. The select is used to choose a tag population based on a value or multiple values in a tag's or multiple tags' memory. The inventory is used to identify individual tags. The reader will initiate the communication by sending queries to tags. Tags will reply to the reader with the requested information. The access is used to communicate with tags, which include reading, writing, locking, and killing the tag.

From the network level, one of the most challenging problems for RFID is collision avoidance. When a reader sends queries to tags, if more than one tag responds, there is a collision and the reader is not able to correctly receive the information. To avoid collisions, anti-collision protocols were designed to efficiently solve this problem.

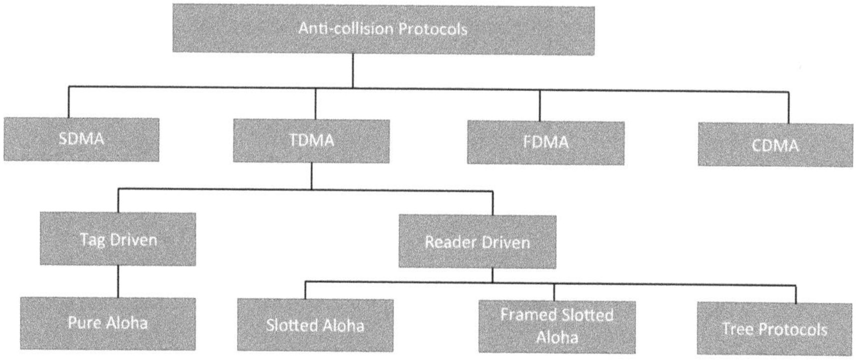

Figure 7.6 Collision avoidance protocols

There are mainly four categories of anti-collision protocols [128]. First, the SDMA (spatial division multiple access) relies on beamforming or directional transmission techniques to reduce the interference from tags. It requires complicated architectures of the transmitter. The TDMA (time division multiple access) allows tags to respond to the reader's queries in sequence to avoid simultaneous transmissions. It reduces the overall throughput but is simple, which is an important factor to design RFID tags due to their limited resources. The FDMA (frequency division multiple access) uses multiple frequency bands to communicate. Since the cross-band interference can be neglected, FDMA allows tags to operate at different frequency bands to avoid interference. The transmitter using FDMA uses complicated circuits and algorithms to demodulate signals. The CDMA (code division multiple access) occupies a broad frequency band and uses a unique code to avoid interference. Its power consumption is relatively high. As a result, TDMA is a practical solution for RFID considering its applications and cost. As shown in Fig. 7.6, most of the anti-collision protocols are designed based on TDMA. A more detailed survey is provided in [128].

7.4 Magnetic Waveguide Network Protocol

We have introduced point-to-point and point-to-multipoint network protocols. For multipoint-to-multipoint network protocol, we can follow existing wireless network stacks. A key challenge to enable multipoint-to-multipoint network using MC is the communication range. Long-range MC can be achieved using magnetic waveguide, metamaterials, ferrite cores, or large coils. There is another approach by using distributed coils to form magnetic waveguides to extend the communication range. In other words, sensors with coils are randomly distributed. When a source sends packets to its destination, which cannot be achieved by using a single hop transmission, distributed scheduling will be performed to find relay coils to form a magnetic waveguide to send packets to the destination in one hop or a small number of hops.

In electromagnetic wave-based wireless networks, multihop relay has been utilized to increase the communication range and improve the received signal quality.

Although this method appears to be promising [131], the end-to-end delay from the data source to the destination is greatly increased. As an illustration, a packet is transmitted from node N_1 to N_n via relays N_2 to N_{n-1}. Assume that each hop takes T_0, in which case $(N_n - 1)T_0$ would be the smallest end-to-end transmission delay without interference from other simultaneous transmissions. The full-duplex relay network has just recently been created. There are still numerous issues that need to be resolved, though, because the self-interference problem has not yet been fully mitigated [132].

Passive relays can be used in MC to create a waveguide, as covered in earlier chapters. The foundations of magnetic relay scheduling are then explained using the example of metamaterial-enhanced MC. Due to the inductive coupling used for communication, MC networks can accomplish full-duplex multihop relay without being affected by self-interference. So long as there are enough relay nodes in between and the waveguide loss is small enough, the source node and destination node can be placed arbitrarily far away from one another. The latency will be T_0 rather than $(N_n - 1)T_0$ in the situation above because all the relays are full duplex. In a full-duplex MC network, the end-to-end delay of a multihop transmission can be as little as one transmission duty cycle. Each node in such a network can connect to any other node with just one hop, considerably enhancing the robustness and efficiency of the network.

It is crucial to comprehend the key distinctions between the electromagnetic wave-based relay and the MC relay. Wireless devices are situated far apart in electromagnetic wave-based communication. The relay can overhear packets sent by the transmitter to the receiver because electromagnetic waves can generate currents in the relay antenna. This generated current can be thought of as a source of re-radiation that broadcasts the signal received by the relay into space. However, for two reasons, the receiver is unable to pick up this signal. First, because the distance between the transmitter and the relay is great in comparison to the wavelength, the induced current in the relay antenna is weak. Additionally, the relay occurs at some distance from the receiver. Because of this, the relay must send the packet again after receiving it because the re-radiated signal cannot be picked up by the receiver. Since a full-duplex electromagnetic wave-based relay transmits and receives data simultaneously, self-interference is a significant problem, and performance is typically sacrificed [132, 133].

MC uses near-field coupling, which is independent of propagating waves, as opposed to electromagnetic wave-based communication [21, 23, 51, 134]. The information can be transmitted via the coupling as a result, which gives the MC transceivers much stronger couplings (high mutual inductance among coils). Due to the near-field coupling, the MC relay is fundamentally full duplex in contrast to the electromagnetic wave-based relay. If a magnetic communication relay is placed between a transmitter and a receiver, the sent signal can cause currents to flow through the relay, much like it would if electromagnetic waves were being used. The relay with induced current can be thought of as a new radiation source, which is the difference. Strong coupling between the receiver and the relay allows it to additionally induce currents at the receiver that carries the sent signal. Due to the relatively limited communication range and strong magnetic coupling, all of the aforementioned activities take place concurrently. A packet is sent by the transmitter in an electromagnetic wave-based half-duplex relay, which is then received by the relay and transferred to the receiving node in the following time slot. As a result, two time slots are used during the transmission procedure. In contrast, only one time slot is needed in full-duplex MC networks to finish the data transmission. Due to the initial distance separating them, the transmitter and the receiver are unable to converse immediately. The coupling between the three nodes may now be strengthened thanks to the full-duplex relay, and the packet can now be sent directly to the receiver.

Then, using a network protocol for relay scheduling as suggested in [135], we take into account an MC network with randomly placed subsurface sensors. In 2D network analysis, the cell partition is frequently utilized [136]. By taking into account an ad hoc network with uniformly placed nodes in a 3D cube with edge length l_s, we extend it to 3D here. The position of node i is indicated by X_i, and the node density is λ_n. We also separate the space into tiny subcubes with edges of l_{sc} length. It should be noted that the setting needs not to be subterranean; it can be any number of different harsh and challenging locations. This cubic structure has numerous practical applications. For instance, in [137], the MI sensors are installed in boxes on a food truck to track the quality of the food in real time; in underwater fish farming, fishes are raised in cages that are set up similarly to the network structure proposed; the MI sensors/robots can be installed in the cage to gather useful data.

For interference from neighboring nodes, the widely used protocol model [138] is taken into account. Assume that $d_{\text{max}}[i]$, where $i = 1,2,\ldots,N_{\text{tap}}$, is the communication range associated with gain $\alpha_{\text{gain}}[i]$. Only if all of the potential interference is $(1 + \Delta)d_{\text{max}}[i]$ distant from the receiver, a link can reach a data rate of $\mathcal{R}$. The shortest route between the source and the destination is used in the routing strategy. All of the nodes in a cell's neighboring cells can communicate with the nodes in that cell. If there are no empty cells, the line linking them serves as the path from the source to the destination. It is possible to configure the node density, cell size, and communication range to ensure that every cell contains at least one node [138–140]. Due to the very low communication range in MC networks, which necessitates the placement of a very high density of nodes in a very compact space, this criterion is difficult to satisfy. Since vacant cells are permitted, although the network is connected, this possibility is taken into account. Each node in the routing algorithm looks at its neighbors and selects the one that is not empty and closest to the destination as the relay. The center of the cells is utilized to calculate the distance between two cells, causing the routing algorithm to converge.

Even if there is barely any delay for a single transmission in the network, this is accomplished by equivalently expanding the transmission range and decreasing the network capacity. There is a trade-off between network capacity and latency, as stated in [136, 140]. While a long communication range can shorten a communication link's latency, it may lengthen the time for other packets waiting to be transferred and slow down the overall network throughput. Short communication range offers a high network capacity but comparatively long delay. The scheduling of relays is then done using a useful MAC protocol.

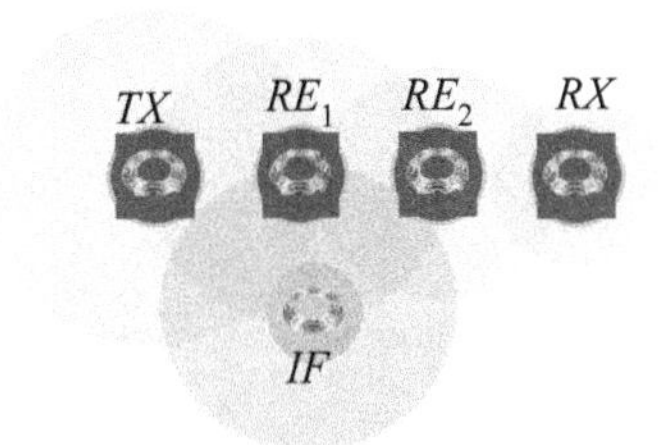

Figure 7.7 Interference in magnetic communication networks using relays. The disk denotes the achievable communication range. TX denotes the source, RX denotes the destination, RE denotes a relay, and IF denotes an interference source.

Assume for the moment that the nodes are omniscient and possess perfect topological knowledge. The two main components of data communication are random medium access and data transmission. In contrast to the data packet, which requires significantly more time, the medium access uses very brief time intervals that can be ignored, as shown in Fig. 7.7. The source S_1 sends an RTS (Request to Send) message to the next-hop cell on the path in the event that the channel is available. A node RE_1 will willingly act as the relay in the following cell, adjusting its metamaterial gain in accordance with the received power. A node named RE_2 will voluntarily act as the relay once RE_1 delivers RTS to the next-hop cell on the path. This process will keep running in a cell until at least one of the following occurs: The destination is within the cell, other transmissions are interfering with the cell's nodes, one node in the cell is relaying packets for other transmissions, and the maximum number of relays has been reached. Since very long distance transmission might significantly affect network capacity, the maximum relay number is taken into consideration to prevent it. However, the network layer can permit a high maximum number of relays if an emergent event is occurring and it is given high priority.

It can be the destination or a relay in a cell x if one of the four conditions is met there. The cell will first wait $(N_{\text{rmax}}+1-x)(T_{\text{RTS}}+T_{\text{CTS}})$, where N_{rmax} is the maximum number of relays, x is the number of hops from the source, and $T_{\text{RTS/CTS}}$ is the time needed to send RTS or CTS (Clear to Send), before sending its CTS if it has the destination D_1. The node RE_{x-1} will send this information to the node before it, once it receives the CTS from RE_x, and so on. All other relays and the destination are ready to send the packet when S_1 receives the CTS and begins transmitting. Second, RE_x will immediately transmit a CTS to the previous relay rather than broadcasting an RTS if the cell x has a relay and the counter approaches the maximum number of relays. In this case, the last relay is regarded as a temporary destination and is not included in the total number of relays. At most $(N_{\text{rmax}} + 1)$ RTS will be sent during the procedure. Since each RTS has a corresponding CTS, $2N_{\text{rmax}} + 2$ time slots are set aside in the protocol for RTS and CTS. The scheduling process will halt at the final node, sending RTS if any RTS are lost as a result of collisions or interference. Thanks to RTS's scheduling system, CTS won't be hampered.

In Fig. 7.8, an illustration is provided. Assuming there can be a maximum of six relays, N_1 represents the source, and N_8 represents a relay or destination. Prior to

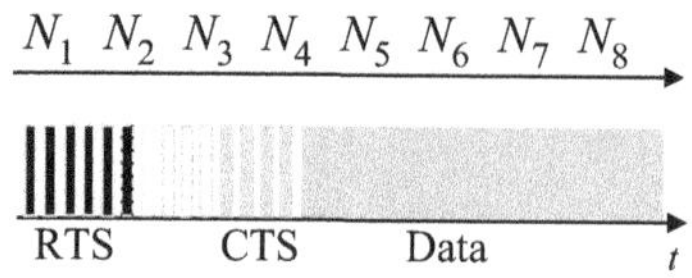

Figure 7.8 Modified CSMA/CA for magnetic communication network

sending a packet to N_8, N_1 sends an RTS to N_2. Take into account that N_2 to N_5 are accessible, but N_6 is being obstructed by other transmissions. So, if N_5 sends an RTS to N_6, N_6 won't reply. Then, N_5 waits for five vacant time slots, two RTS and three CTS, before sending N_4 a CTS. Until N_1 receives a CTS from N_2, this process will continue. The data packet is then transmitted, and N_2 to N_4 will act as full-duplex relays, with N_5 acting as the receiver.

7.5 Summary

Depending on the specific application, MC networks have various stack protocols. In this chapter, we divide the applications into three categories on the basis of the number of sources and destinations. MC is unique since it allows extremely short-range communication that can only support one transmitter and one receiver without creating interference for other users. NFC and RFID are among the most widely used MC protocols on which several industrial standards have been developed. The multipoint-to-multipoint MC network is newly developed thanks to the advancement in long-range MC. Its networking protocols are still under development.

Part III

Applications

8 Applications of Magnetic Communications Systems

In this chapter, several kinds of MI-based applications are introduced, specifically, the MI-based localization system that is one of the most widely used and mature applications of the MI-based techniques. Thus, this chapter first describes several typical MI localization applications that are the motion capture system, pipeline position systems, and fusion localization with other techniques (such as inertial measurement correction). Second, we summarize some MI-based communication applications for IoT, such as RFID, through-the-earth (TTE) communication, and underwater communication.

8.1 MI-Based Localization

8.1.1 Overview of MI-Based Localization

The MI-based signal with the wavelength from 10 to 100 km is hardly reflected and refracted in the near-field range. Such a phenomenon caused that the multipathing effect does not happen in the MI-based communication. Therefore, the MI-based signal can not only be used for information interaction (communication) but also for localization in the challenging environment, such as underground and underwater scenarios.

In 1979, Raab et al. proposed the approach of magnetic-based localization and tracking (MLT) [141]. In [141], they study the relationship between the magnetic coupling and object position in the position system that is composed of the source of 3-axis dipole, a triaxial sensor, and their (the 3-axis dipole and triaxial sensor) supporting electronic circuits. This research can be treated as the primitive theoretical guidance for underground localization. Subsequently, their MLT approach is improved by E. Paperno at Ben Gurion University [142]. In their studies, two mutually orthogonal coils fed with phase-quadrature currents comprise the excitation source, which is equal to a mechanically rotating magnetic dipole. The resulting excitation field rotates elliptically at any position in the near-field region. The AC part of the squared field magnitude is a sinusoidal wave which is at twice of the excitation frequency. The following set of parameters uniquely characterizes the excitation at the sensor's position. Three years later, on the basis of such a technique, Paperno et al. further studied the biaxial sensor that is tracked with a triaxial transmitter [143]. Compared to the uniaxial sensor, their biaxial sensor can get the position of an object and tracking information with six degrees of freedom by using the non-iterative algorithm. In 2013, at Ben Gurion University also, Sheinker's team studied the 2-D MI-based localization methods using the beacon of a low-frequency quasi-static magnetic field [144]. The different magnetic fields are generated by configuring the different beacons. The fields are induced by the array of the biaxial sensors and produce the signals for object localization. Furthermore, they also use such a method for the 3-D MI-based localization by replacing the biaxial sensors with the triaxial sensors [145].

Furthermore, some researchers focus on more specific applications. For example, Markham et al. at Kellogg College studied the 3-D MI-based localization for the underground animals [146]. In this study, animals worn a collar that could record the magnetic field strength and upload the magnetic signal data to the host. Thus, the position and tracking of the animal could be calculated using these uploaded information. Markham's research is very useful for learning the habits of underground animals. In 2015, X. Tan and Z. Sun considered the effect of the surrounding high permeability material on the MI-based localization.

The MI-based localization can also be used in the medical area. After Meng et al. at Hong Kong University of Science and Technology proposed the wireless robotic capsule endoscopy and the internal structure [147], C. Hu and team at Ningbo University studied the position and orientation of the endoscopy in the human body [148]. They develop the endoscopy system based on the array of 3-axis magnetic sensors to track the position and orientation of the magnet body. Since the magnetic strength suffers significant attenuation with the increasing distance between the magnet and sensor, B. Li at Chinese Academy of Sciences developed the system of magnetic localization and orientation for capsule endoscope with weak magnetic signals [149].

From above researches, a remarkably different characteristic between the MI-based localization and communication can be found. For the MI-based localization applications, to distinguish between different locations and simplify the localization system, the difference of signal among different points is expected to be as significant as possible. Conversely, for MI-based communication applications, the difference of signal among different points is expected to be small since it can not only simplify the hardware and software architecture but also decrease the energy consumption of the communication system.

8.1.2 Typical Applications

Magnetic Motion Capture System

Motion capture (MoCap) is the key part of moving object tracking. The magnetic MoCap is a MoCap technique using magnetic signals. Compared with the optical MoCap (such as the camera monitor system), the magnetic MoCap is useful to detect hidden objects, and it provides the low-cost systems, such as application for geophysics, vehicle detection, detection of one dipole.

The MoCap system is usually composite of several markers and magnetic field sensor. The marker can be installed on the target motion object. It can generate the magnetic fields that are sensed by the sensors and then calculates the position,

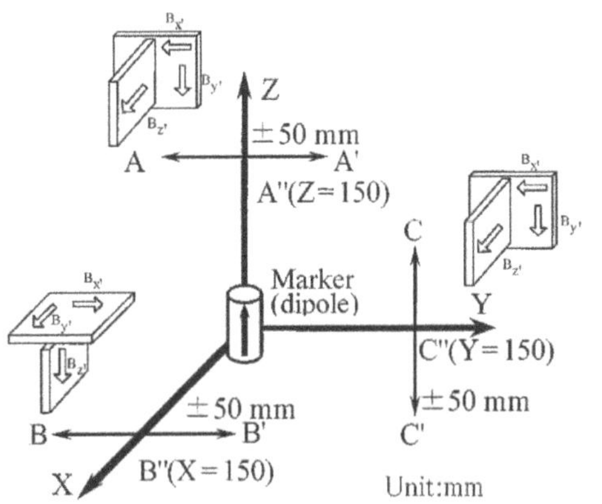

Figure 8.1 Arrangement sensors and a marker [150]

orientation, and moving state of the markers. For example, Yabukami et al. at Tohoku University developed a MoCap for the marker with a magnetic dipole and several markers with a 3-axial magnetic field sensor [150]. Their hardware architecture of the MoCap system is as shown in Fig. 8.1. The magnetic marker in Fig. 8.1 is composed of cylindrical NdCFeCB magnet with 5 mm diameter and 10 mm length. The magnetic marker is magnetized along the longitudinal axis with the magnet whose magnetic moment is 2.3×10^{-7} Wbm. In Fig. 8.1, sensors are set on holding plates and markers are held by three-axial micrometer, both are made of aluminum (non-magnetic). Therefore, Magnetic flux density $\vec{B}_{\text{cal}}$ around a magnetic dipole source is calculated by

$$\vec{B}_{\text{cal}} = \frac{1}{4\pi}\left(-\frac{\vec{M}}{|\mathbf{r}|^3} + \frac{3(\vec{M}\cdot\mathbf{r})\mathbf{r}}{|\mathbf{r}|^5}\right), \tag{8.1}$$

where $\vec{M}$ is the magnetic moment of the marker, $\mathbf{r}$ is the position vector of the marker. The position and orientation of dipole were obtained by solving

$$\vec{B}_{\text{cal}} - B_m = 0. \tag{8.2}$$

Equation (8.2) consists of 12 equations if the number of markers is 2. Equation (8.2) consists of six equations if the marker number is 1. These equations can be calculated by the Powell method.

As shown in Fig. 8.1, when the sensor sweeps along $A - A'$, $B - B'$, and $C - C'$, and the three components of the sensor are depicted in each sweep direction, the position accuracy can be calculated as

$$dl = |\mathbf{r}_{\text{means}} - \mathbf{r}_{\text{set}}|, \tag{8.3}$$

where $\mathbf{r}_{\text{means}}$ is the position calculated by (8.2) and $\mathbf{r}_{\text{set}}$ is the actual position.

MI Localization in Pipeline Environment

Pipelines provide the most economical way to transport large quantities of fluids, such as crude oil, natural gas, and sewage. Since pipelines are often buried under the ground, it is very difficult and expensive to obtain the in situ and real-time information

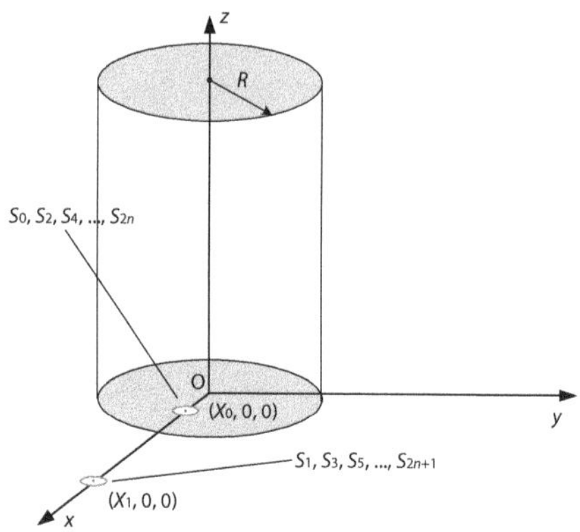

Figure 8.2 Positions of the images in straight pipeline [151]

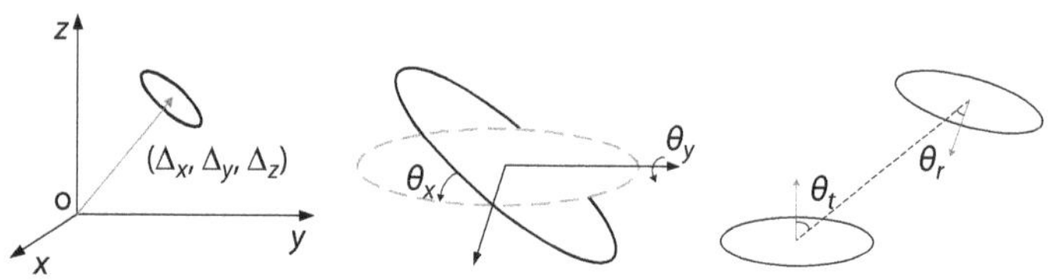

Figure 8.3 Coil position and rotation in Cartesian coordinates [151]. (**a**) Position of the coil, (**b**) rotation of the coil, and (**c**) definition of θ_t, θ_r

inside the pipelines. In addition, depending on the applications, the pipelines may be filled with water, oil, or other fluids, which may prevent the efficient propagation of EM signals. Therefore, the MI-based localization is very effective. For example, Tan et al. studied the MI localization in the straight line and pipeline network [151]. As shown in Fig. 8.2, a coil is located at $(x_0,0,0)$ with the rotation angles $\theta_x = 0°$ and $\theta_y = 0°$. Here, θ_x and θ_y, respectively, indicate the angle rotation around the x axis and the y axis from the horizontal, as shown in Fig. 8.3(b). The reflected magnetic field by the metallic pipe wall can be regarded as a field generated by a new source outside the pipeline. The position of the new source can be expressed as $x_1 = \frac{R^2}{x_0}$, where R is the radius of the pipeline. By considering multiple reflections, the new source outside the pipeline will take a second image inside the pipeline superposed on $(x_0,0,0)$. Similarly, this second image will take a third image outside the pipeline. Therefore, the images including the original source inside the pipelines are $s_0, s_2, s_4, \ldots, s_{2n}$, and the images outside the pipeline are $s_1, s_3, s_5, \ldots, s_{2n+1}$. The magnetic field generated by nth image is expressed by

$$\mathbf{H}_n(x,y,z) \simeq \frac{C_r^n N I_0, r^2}{4} \cdot \left[\frac{3a_n c_n}{(a_n^2 + b_n^2 + c_n^2)^{\frac{5}{2}}} \mathbf{i} + \frac{3b_n c_n}{(a_n^2 + b_n^2 + c_n^2)^{\frac{5}{2}}} \mathbf{j} + \frac{2c_n^2 - a_n^2 + c_n^2}{(a_n^2 + b_n^2 + c_n^2)^{\frac{5}{2}}} \mathbf{k} \right] \cdot e^{-j\omega(t-\tau_n)}, \tag{8.4}$$

where $C_r = \frac{\mu_1-\mu_2}{\mu_1+\mu_2}$, μ_1 is the permeability inside the pipeline, μ_2 is the permeability of the pipe wall, a_n, b_n, and c_n are given by

$$a_n = \begin{cases} x - \dfrac{R^2}{x_0} & \text{if } n \text{ is odd,} \\ x - x_0 & \text{if } n \text{ is even,} \end{cases} \tag{8.5}$$
$$b_n = y,$$
$$c_n = z,$$

and τ_n is the phase shift caused by the signal propagation time and is expressed as

$$\tau_n = \begin{cases} \dfrac{(-2Rn - 2R + x_0 - x_r)^2 + (y_r - y_0)^2 + (z_r - z_0)^2}{c} & \text{if } n \text{ is odd,} \\ \dfrac{(2Rn - 2R - x_0 - x_r)^2 + (y_r - y_0)^2 + (z_r - z_0)^2}{c} & \text{if } n \text{ is even.} \end{cases} \tag{8.6}$$

Hence, the superposed magnetic field generated by the source and infinite images can be calculated by

$$\mathbf{H}(x,y,z) = \sum_{n=0}^{\infty} \mathbf{H_n}(x,y,z). \tag{8.7}$$

For the pipeline with the turns and junctions, as shown in 8.4, Node 1 is located at $(x_0, 0, 0)$ in Pipeline 1 working as the transmitter. The receiving node is located at $(0, y_r, \Delta L)$ in Pipeline 2. There are two portions of magnetic field components at the receiving node that can be calculated independently. For the first portion, by neglecting the influence of Pipeline 2, Node 2 is supposed to be deployed outside the Pipeline 1. Therefore, the case becomes the same as it is in single straight pipelines, and hence the magnetic field generated by the transmitting coil can be calculated using (8.7) and (8.17). For the second portion, the magnetic field reflected by Pipeline 2 can be expressed as

$$\mathbf{H}_n^{\clubsuit}(x,y,z) \simeq \frac{C_r^n N I_0, r^2}{4} \cdot \left[\frac{3a_n^{\clubsuit} c_n^{\clubsuit}}{(a_n^{\clubsuit 2} + b_n^{\clubsuit 2} + c_n^{\clubsuit 2})^{5/2}} \mathbf{i} + \frac{3b_n^{\clubsuit} c_n^{\clubsuit}}{(a_n^{\clubsuit 2} + b_n^{\clubsuit 2} + c_n^{\clubsuit 2})^{5/2}} \mathbf{j} + \frac{2c_n^{\clubsuit 2} - a_n^{\clubsuit 2} + c_n^{\clubsuit 2}}{(a_n^{\clubsuit 2} + b_n^{\clubsuit 2} + c_n^{\clubsuit 2})^{5/2}} \mathbf{k} \right] \cdot e^{-j\omega(t-\tau_n^{\clubsuit})}, \tag{8.8}$$

$$c_n^{\clubsuit} = \begin{cases} z - \dfrac{\Delta L^2 - R^2}{\Delta L} & \text{if } n \text{ is odd,} \\ z & \text{if } n \text{ is even,} \end{cases} \tag{8.9}$$
$$a_n^{\clubsuit} = x - x_0,$$
$$b_n^{\clubsuit} = y,$$

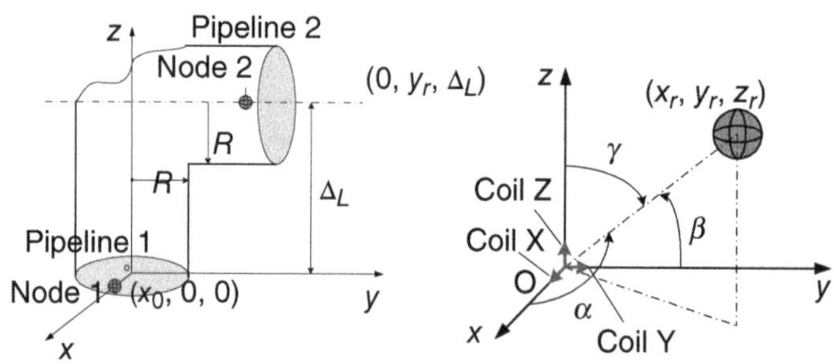

Figure 8.4 Coil position and rotation in Cartesian coordinates [151]. (**a**) Nodes in different pipes and (**b**) the relative position between two sensor nodes

$$\tau_n^{\clubsuit} = \begin{cases} \dfrac{(z_0 + \frac{3}{2}nR - \frac{7}{2}R - 2\Delta L - z_r)^2 + (y_r - y_0)^2 + (x_r - x_0)^2}{c} & \text{if } n \text{ is odd,} \\ \dfrac{(4\Delta L - \frac{3}{2}nR - z_0 - z_r)^2 + (y_r - y_0)^2 + (x_r - x_0)^2}{c} & \text{if } n \text{ is even.} \end{cases} \tag{8.10}$$

Therefore, the total magnetic field at the receiver for two pipeline scenarios is

$$\mathbf{H}(x,y,z) = \sum_{n=0}^{\infty} \mathbf{H_n}(x,y,z) + \sum_{n=0}^{\infty} \mathbf{H_n^{\clubsuit}}(x,y,z). \tag{8.11}$$

Once we obtain the relationship between the magnetic field $\mathbf{H}$ and the coordinate (x,y,z) of the receive coil, then we can obtain the magnetic flux with respect to the injected current I_0, which is $M(x,y,z) = \frac{\partial(\pi\mu_1 r^2 |\mathbf{H}(x,y,z)\cdot\mathbf{n}|)}{\partial I_0}$. The ratio of the received power to the transmitted power in MI communication can be expressed as

$$h^2(x,y,z) = \frac{P_r}{P_t} \simeq \frac{\omega^2 M^2(x,y,z)}{2R_0^2 + 2\omega^2 M^2(x,y,z)}, \tag{8.12}$$

where R_0 is the unit-length resistance of the coil and ω_0 is the operating frequency.

As shown in Fig. 8.4(b), the received signal strength (RSS) coming from the Coils X, Y, and Z are P_{rX}, P_{rY}, and P_{rZ}, respectively. The total RSS coming from the three transmitting coils is denoted as P_r. From Fig. 8.4(b), we know $x = d\cos\alpha$, $y = d\cos\beta$, and $z = d\cos\gamma$, where d can be expressed by (x,y,z) and the pipe radius by R using elementary geometry. Also, d can be obtained by solving Eq. (8.12) if we measure the power of the transmitter P_t and the power of the receiver P_r. Hence, the coordinate estimator is

$$\hat{x} = \frac{P_{rX}}{P_r}\left(\frac{3}{2}\frac{\eta K P_t}{P_r}\right)^{1/6}, \quad \hat{y} = \frac{P_{rY}}{P_r}\left(\frac{3}{2}\frac{\eta K P_t}{P_r}\right)^{1/6}, \quad \hat{z} = \frac{P_{rZ}}{P_r}\left(\frac{3}{2}\frac{\eta K P_t}{P_r}\right)^{1/6}, \tag{8.13}$$

where

$$K = \frac{\mu_1^2 \omega^2 N^2 r^6}{16R^2 + \mu_1^2 \omega^2 N^2},$$

$$\eta = \begin{cases} \dfrac{1 + \frac{x_0^2}{R^2} C_r^2}{1 - C_r^4} & \text{single pipe} \\[2ex] \dfrac{1 + \frac{x_0^2}{R^2} C_r^2}{1 - C_r^4} + \dfrac{\frac{\Delta L^2}{R^2} C_r^2 + \frac{x_0^2}{R^2} C_r^4}{1 - C_r^4} & \text{two pipes.} \end{cases} \quad (8.14)$$

Sub-meter MI Localization with IMU Fusion

In the MI localization system using the magnetic field generated by the MI base stations (transmitter), the cumulative error would increase with an increase in the displacement of the moving object. Also, for some heavily distorted environments, there is a significant decrease in accuracy. In 2015, Abrudan et al. at University of Oxford developed a high-accuracy MI-based localization system with a multi-level fusion approach [12], using an inertial measurement unit (IMU). The architectural design and topology of this fusion localization system are shown in Figs. 8.5 and 8.6, respectively.

The fusion localization system comprises a triaxial transmitter (TX) and triaxial mobile receiver (RX). The transmitter with three orthogonal coils generates low-frequency magneto-inductive fields, which can be arbitrarily oriented in 3-D space. The mobile receiver has triaxial receiving coils that detect the weak quasi-static magnetic fields and determine accurate 3-D positions with the aid of an IMU.

As shown in Fig. 8.6, the architecture of the fusion localization system is a composite of a sensor block, an MI signal processing block, and a localization block. The sensor block has three mutually orthogonal coils that act as MI sensors measuring a 3-D-modulated magnetic vector field in a particular location.

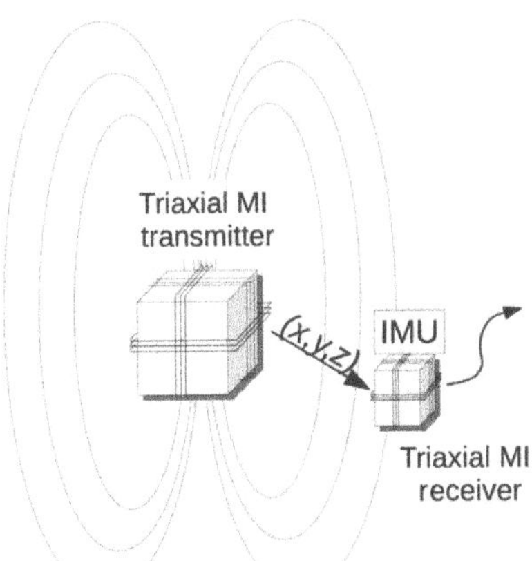

Figure 8.5 The MagLoc system: A triaxial transmitter generates low-frequency magneto-inductive fields that can be arbitrarily oriented. A mobile MI receiver detects these fields and with the aid of an IMU determines accurate 3-D positions [12]

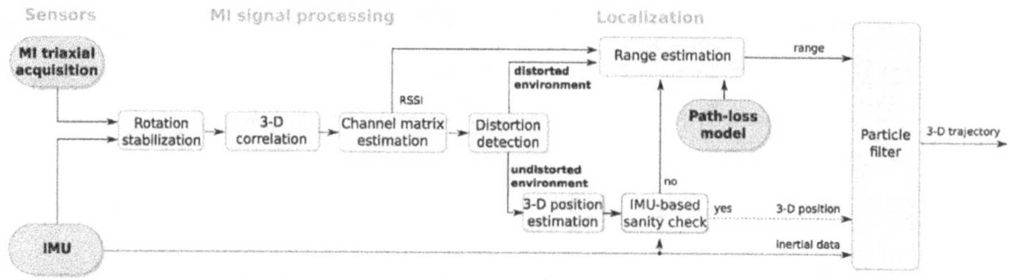

Figure 8.6 Receiver architecture design, showing how MI and IMU data are processed in order to achieve real-time 3-D tracking [12]

The sensor block also has an IMU for the three purposes: (1) Providing a stable RX navigation frame to avoid the effect of the fast changes in the user orientation during the transmission of a frame. (2) Performing sanity check of the MI-based estimated 3-D position displacements over a short time period. (3) Tracking the receiver position over longer time periods using a particle filter. Sensor fusion is done at three different functional layers, not just in the final filtering stage.

In the MI signal processing block, MI signals are passed through a correlator that uses short, low power codes for short-range navigation and long, high power codes for long-distance positioning. Based on the correlator output, the channel matrix is estimated along with the corresponding magnetic receiver signal strength indicator (RSSI).

In the localization block, the distortion detection sub-block automatically determines whether the environment acts similarly to free space or if it acts to distort the magneto-inductive field. This property allows us to use sporadic non-distorted 3-D position estimates to calibrate and compensate the system.

8.2 MI-Based Internet of Things

Although IoT has not a standard architecture, its universally acknowledged architecture is shown in Fig. 8.7.

As mentioned in above chapters, since the magnetic signals can penetrate most nonferromagnetic material and the limited interference created to the adjacent transceivers, the MI-based technique might be a good candidate for the upcoming IoT application for underground, underwater, and other RF challenge or short-range communication scenarios. In fact, most studies on the MI-based communication and applications focus on the perception layer and network layer in IoT architecture shown in Fig. 8.7 (i.e., the MI-based wireless sensor network (WSN)).

8.2.1 Very Short Range Applications

The NFC is a typical RF identification application [152]. It is also a typical very short range MI-based application with transmission ranges below 4 cm. Since the NFC technique is the most mature of the MI-based techniques, the NFC has been widely used in

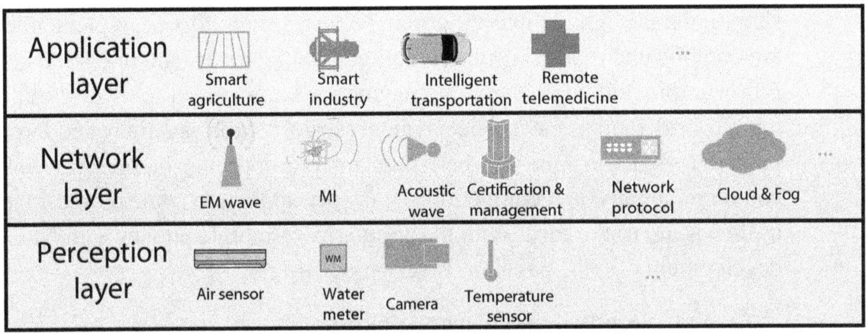

Figure 8.7 The universally acknowledged architecture of IoT

contactless payment and keycards. Here, the short-range transmission turned out to be an advantage of the MI transceivers, since it reduces the risk of eavesdropping and the interference to adjacent communication links. The channel model and the corresponding near-field effects have been considered in multiple work. In the meantime, an NFC forum has been established, where the recent advances of the NFC and its standardization are presented. The achievable data rate of an NFC point-to-point link is only 424 kbit/s according to the existing standards of ISO/IEC 18092/21481, which are based on the passive RFID standard ISO/IEC 14443. This data rate is mostly dictated by the choice of passive circuit elements and few practical assumptions [152].

In order to be compatible with the contactless card, the standard of the NFC has three working modes, i.e., point-to-point communication (PPC) mode, read and write (RW) mode, and the mode of NFC card simulation (NCS).

For the PPC mode, the NFC device exchanges the data with the equivalent identity. For example, several digital cameras and mobile phones with the NFC communication models are interconnected using wireless NFC techniques to exchange virtual cards, digital photos, and other data. The key characteristic of the PPC mode is that each NFC model exchanging data is active (i.e., has a power supply).

For the RW mode, the NFC device is treated as a contactless reader. Different from the NCS mode, the NFC device works as a master that is similar to the inter-integrated circuit (I^2C) master, serial peripheral interface master, etc. For example, the mobile phone with an NFC model can read data in the electronic tag.

For the NFC mode, the device (such as the mobile phone) with the NFC model is disguised as a contactless electronic tag such as the access card and bank card. Such a device can be looked on as a slave device.

By above three modes, the NFC is widely used in our lives. There are three basic types of applications of NFC:

(1) NFC Payment.

The NFC payment is defined as the virtualization of the NFC model in the mobile phone as a payment card, such as the bank card, coupon, and discount card of supermarket. Although NFC is a mature technique with few interferences from environments, it is an open-loop application. Ideally, a mobile phone with an NFC model can be simulated as a bank card that can pay money on the POS of all supermarkets, shopping malls, etc. However, it may be very difficult to be implemented since there are complicated industrial chains, commercial interests, and industrial pattern games. For example, although electronic payment has thrived in the world that has hatched many world-famous electronic payment services, such as Alipay and PayPal, the NFC payment seems to miss the best time for the booming mobile payment due to the commercial game and competition. It is only applied in some closed-loop areas, such as the virtual traffic card. With the popularity of mobile phones with NFC support, the development of NFC payment is very promising.

(2) NFC Security Surveillance Applications

In the security surveillance applications, the mobile phone with an NFC model is simulated as access cards, electronic tickets, and electronic passes. Thus, there is no need to use a smart card. We can achieve access control functions, the configuration, monitoring, and modification of access control using offline and online services. We can use these NFC applications for temporary credentials distribution. In the application of NFC virtual electronic tickets, the ticketing system will send ticket information to the mobile phone. The mobile phone with the NFC function can simulate an electronic ticket with such information. Using a mobile phone as a virtual card, people can reduce the use of the access card or magnetic card ticket, and significantly reduce the cost of security surveillance, and achieve the degree of automation, personnel cost reduction, and efficiency improvement.

(3) NFC Tag Application

In NFC tag applications, after writing posters, promotional information, and commercial advertising into the NFC tag, users can subscribe their interested information by the NFC model. It can also be used to login the social forum or share the information with the world. However, with the development of QR code with more low cost and increasing bandwidth of the wireless network, the development of the NFC tag applications is not promising.

8.2.2 MI-Based Underwater IoT Applications

It is known that water covers more than 70% of Earth's surface in the form of oceans, river, and lake. Especially, sea is an important space for human's survival and sustainable development. There are a variety of natural resources, such as submarine oil, gas, and mineral deposits. However, up to 95% of the sea is still unexplored or insufficiently developed so far [153]. The Internet of Underwater Things (IoUWT) is considered as one of the revolutionary technologies not only in the case of ocean exploration but for the maintenance of its biodiversity as well [154].

With the recent advancements in technology, reliable electronic devices are required to develop practical IoUWT networks. These devices are intended to use any available technology to communicate effectively in the IoUWT network. The architectural aspects of IoUWT and corresponding sensing devices are depicted in IEEE P2413 and IEEE P2510 standards. The basic architecture of IoUWT is shown in Fig. 8.8. From Fig. 8.8, the acoustic technique is used to facilitate communication in the water, while

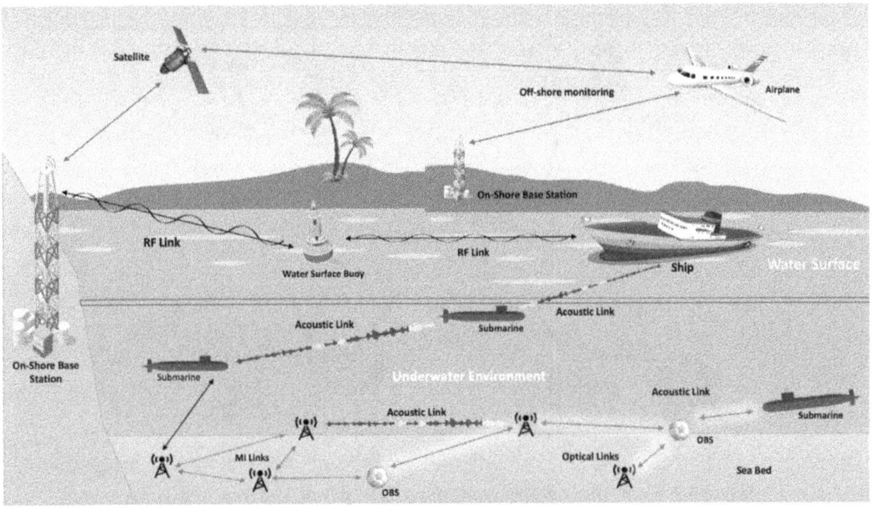

Figure 8.8 Architecture of Internet of Underwater Things [155]

the MI technique is used to facilitate communication in the seabed, since acoustic waves and MI waves suffer less attenuation in the most nonferromagnetic medium.

Acoustic communications are originally developed for military applications with a long history that can date back to the World War II. The acoustic wave can transmit data over a long range (up to several tens of kilometers) with relatively acceptable energy consumption. However, since the acoustic wave speed is much smaller than the EM waves, it has limitations of high transmission delay (usually in seconds) and a relatively low transmission data rate (usually of the order of Kbps). In addition, due to the refection of the acoustic wave, underwater acoustic communication may experience severe multipath fading, if there are large objects, such as large animals, plants, and coral reefs surrounding or existing the path of transmissions. Therefore, MI-based communication is a promising alternative information exchange approach. Compared with the acoustic wave communication, the merits of MI-based communications are not limited to the multipath effects and remarkable signal propagation delay. Therefore, MI communications can be applied into a wide variety of marine applications.

During 2019–2020, Li (at Huazhong University of Science and Technology) and Khalil (at University of Engineering and Technology; see Fig. 8.21) summarized the IoUWT or potential of IoUWT, respectively [153, 154]. The examples of the application of IoUT is shown in Fig. 8.9, and the most widely used applications are introduced below.

(1) Military Exploration

Underwater military applications usually require stealth operations to avoid being detected. Considering the inaudibility and invisibility of underwater MI communications, the MI-based techniques are widely employed to assist military applications, such as information exchange between the autonomous underwater vehicle (AUVs) and submarines.

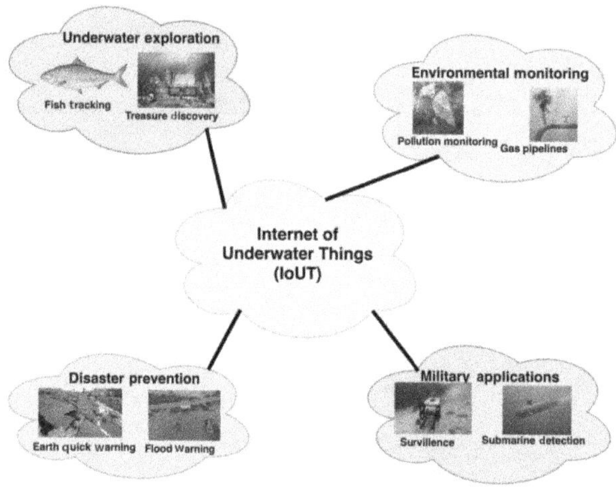

Figure 8.9 Examples of applications of Internet of Underwater Things

(2) MI-Based Disaster Detection

Marine activities are threatened by potential natural disasters, such as typhoons, tsunamis, and undersea earthquakes. Underwater MI communications can be explored to construct timely and robust disaster detection and warning systems.

(3) Underwater Environment Monitoring

Deploying numerous sensors into the sea to form underwater wireless sensor networks (UWSNs) is an important way to monitor the physical, chemical, and biological states of the sea. The implementation and energy costs of MI coils are usually low.

(4) MI-Based Leakage Detection

Accurate leakage detection of underwater oil and gas pipelines is rather challenging due to the hostile and inaccessible underwater environments. It has been shown that deploying high-density MI sensors along the pipelines can provide the necessary measurements for leakage detection with a high resolution but a relatively low cost.

8.2.3 MI-Based Underground IoT Applications

Different from MI-based IoUWT applications introduced above, the acoustic techniques may not be alternative communication methods for the underground communication applications. The reason is that the types of underground materials are much more than underwater ones. The acoustic wave would have significant reflection at the boundary of each material. Thus, the multipath fading of acoustic wave signals would be much more serious than the MI signals. Also, optical techniques are not an alternative communication applications since most of the underground materials are optically opaque. The applications of MI communications in under water environments is summarized in Table 8.1.

Table 8.1 A summary of applications of MI communications in underwater environments

Applications	Scenarios
Military exploration	Surveillance Seabed mine detection Submarine communications
MI-based disaster detection	Typhoon and tsunami detection Land sliding detection Earthquake and volcanic eruption detection
Underwater environment monitoring	Pollution monitoring Mining and transport tunnel condition monitoring Habitat monitoring (marine life, fish farms, reef)
MI-based leakage detection	Underwater oil and gas pipelines Seabed optical fiber cables

In 2006, Akyildiz and Stuntebeck at Georgia Institute of Technology introduced the concept of a wireless underground sensor network (WUSN) that lies in the perception layer and network layer in the universally acknowledged architecture of IoT. They also summarized the application of WUSN and classified current and potential underground applications into four categories: *environmental monitoring, infrastructure monitoring, location determination*, and *border patrol* and *security monitoring* [156]. Twelve years later, Kisseleff also gave a brief discussion for these four categories [152] on the basis of Akyildiz's literature:

(1) Environmental Monitoring

WUSNs have been suggested for agricultural purposes, to monitor the soil condition without creating any disturbance for the aboveground devices, such as tractors and lawn mowers. Here, more precise information can be acquired from an underground system such that a targeted irrigation can be done more efficiently than with existing solutions.

The earthquake and landslide prediction is an important application. For this, the relative movement of the soil layers can be carefully monitored by deeply buried sensors such that possible disasters can be predicted well in advance more reliably than with the existing methods.

The detection of toxic substances that would pollute the environment is an application of WUSNs. To prevent the contamination of drinking water supplies, a thorough monitoring of the chemical content of the soil and water is needed. As a result, a hybrid network has been proposed which would be deployed partially in the underground medium and partially in the underwater medium. The affected regions and their population can be warned and evacuated much earlier. Such a network can also monitor the glacier movements and volcanic eruptions.

(2) Infrastructure Monitoring

The underground infrastructure such as pipes, electrical wiring, and liquid storage tanks can be monitored. Through this, finding the leakage in a pipe or the disconnection of an electric cable would be facilitated. Furthermore, the structural health of underground coal mines is of utmost importance for the safety of mine workers. Hence, sensor nodes need to be deployed in the walls and ceiling of the mine galleries to detect any structural instability.

(3) Security

For the security application, WUSNs can be deployed to detect an intruder and alert the security team. The presence of sensors deployed below the surface can be hardly detected. Correspondingly, the security system is less likely to be discovered and disabled by the intruder. Similarly, WUSNs can be deployed along the border in order to detect illegal crossings and notify the authorities. Here, high precision of the crossing detection using WUSNs may help the authorities to easily localize the intruder.

8.2.4 Through-the-Earth Applications (Very Long Range Applications)

The MI-based TTE communication with tens to hundreds of distance meters can be used in the mines, tunnels, and mountains. In 2011, Lockheed Martin developed the MagneLink^{TM} Magnetic Communication System (MCS) that achieves a two-way communication between underground and the surface. In a test in 2012, Lockheed Martin successfully recorded a longest propagation range of 472.5 m for two-way TTE communications using 8-m transmit antennas. Likewise, a team of Tongji University conducted a two-way wireless TTE communication experiment in the coal mine and Sheshan Mount. In their experiment, a vertical TTE transmitting distance of 310 m was achieved using 8-m transmit antenna [157]. In addition, Vital Alert Company developed the MI-based TTE productions achieving above 450-m communications for the underground tunnel applications.

8.3 MI-Assisted Underwater CPS

8.3.1 MI-Based Underwater Communication System

(1) Underwater Communication Application Scenarios and Problems

In the past few years, people have been paying attention to underwater communication, and this field is full of rich theoretical and practical challenges. Most of the work of underwater communication is carried out on the basis of acoustic communication, and it also includes EM and light waves, which are combined with applications supported by large underwater equipment networks, such as common autonomous driving robots and large sensors. You can obtain the required information from the aquatic or marine environment, and then process the extracted information, and send it to the remote terminal to become the information that people need. However, the current underwater environment poses a huge challenge to the reliability and real-time performance of acoustic communication. Acoustic communication exposes many problems, such as high propagation delay, low data rate, high bit error rate, and channel behavior that are highly dependent on the environment. The same EM wave also exposes the problem of large attenuation underwater, which limits the range of communication

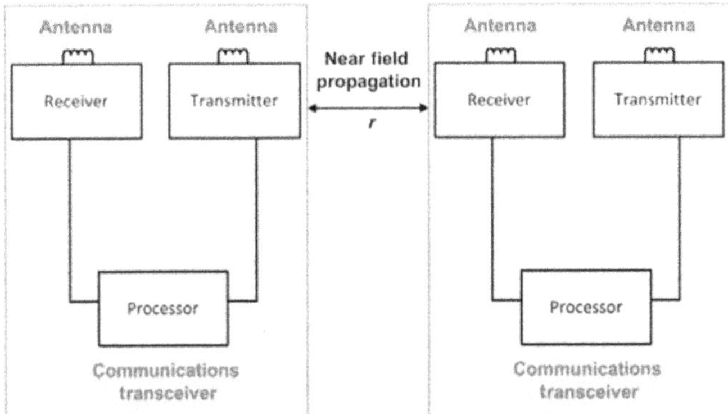

Figure 8.10 Underwater communication system [6]

propagation [158]. The existing solution to this problem is to equip a larger antenna, but it is not suitable for small underwater robots or aircraft.

MI communication can solve the above problems due to its inherent advantages in predictable channel response and competitive energy consumption. It has attracted widespread attention. At the same time, it is a very promising solution due to its small propagation delay and low sensitivity to the surrounding environment. Underwater MI communication uses time-varying magnetic fields to obtain messages from senders and receivers, which just overcomes the drawbacks exposed by traditional underwater communication modes that rely on EM, light waves, and sound waves [6].

(2) Technical Feasibility of Underwater MI Communication

The MI communication system generally runs very close. At this time, the MI communication system is equivalent to a short-range wireless physical layer, which can communicate through a tight, low-power, non-propagating magnetic field between the couplers [6]. At this time, the coil of the transmitter can modulate the coil of the receiver to measure the magnetic field.

In this chapter, it is assumed that the magnetic coupling antennas of the underwater transmitter and the underwater receiver are separated by a distance r, so that the underwater MI communication transmission system, as shown in Fig. 8.10, can be obtained.

The dual-antenna system can be transformed into a dual-port network as shown in Fig. 8.11.

Next, we calculate the path loss value of the MI communication system. In this chapter, it is assumed that none of the antennas in the dual-antenna system is lost. After a series of calculations, through the mutual impedance and self-impedance of the transmitting antenna and the receiving antenna, the mutual inductance between the coils, the self-inductance of the magnet tube, the load impedance at the receiver can be calculated, and then the path loss PL can be obtained:

$$\mathrm{PL_{MI}} = -10\log\left(\frac{R_L\omega^2M^2}{R_{TX}(R_L+R_{RX})^2+R_{TX}(X_L+\omega L_{RX})^2}\right). \tag{8.15}$$

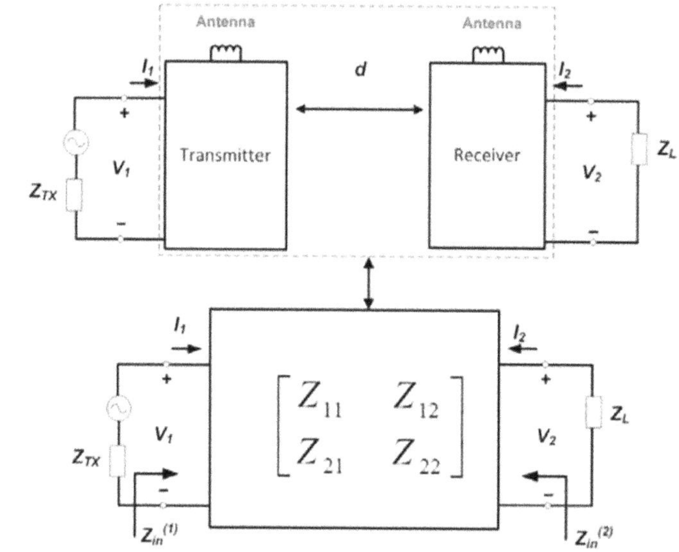

Figure 8.11 Equivalent two-port network of a dual antenna system [6]

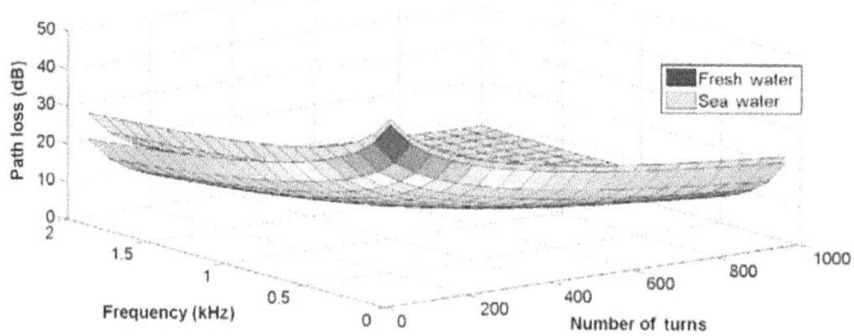

Figure 8.12 Path loss for MI communication as a function of frequency and number of turns [6]

The AC magnetic field can generate eddy currents in salt water and polluted water, thereby generating a negative moment, which offsets the transfer moment and changes the distribution of the field [6]. Therefore, we can obtain the path loss in seawater as

$$\mathrm{PL_{SW}} = \mathrm{PL_{MI}} + \mathrm{PL}_{\alpha}. \tag{8.16}$$

Figure 8.12 shows the path loss in fresh water and seawater with respect to frequency and the number of coil turns. It can be observed that the path loss decreases with the increase in the number of coil turns and frequency, and the path loss in seawater is much larger than that in fresh water [6].

Figure 8.13 shows the path loss in fresh water and seawater with respect to the coil radius and frequency. It can be seen that the path loss decreases with the increase of frequency and coil radius, and for a large coil radius, the dependence between path loss and frequency is lower. Compared with fresh water, the path loss of seawater is still larger [6].

After modeling and theoretical analysis of the underwater MI communication channel, it can be concluded that the freshwater MI system has lower path loss and bit error

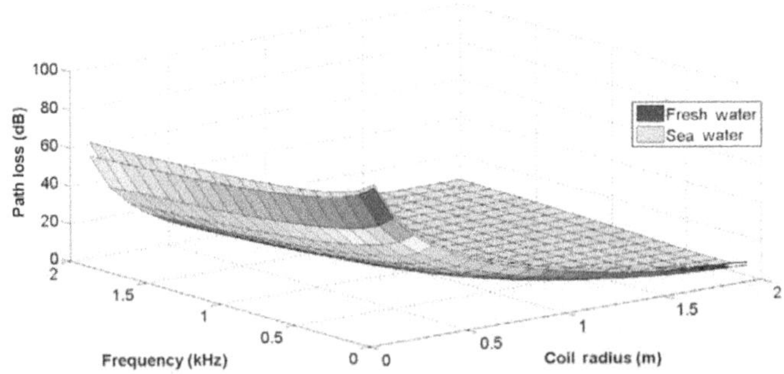

Figure 8.13 Path loss for MI communication as a function of frequency and coil radius [6]

rate than EM and acoustic communication systems during short-distance transmission. Due to the conductivity of sea water, the path loss of the MI system in seawater is higher than that of fresh water, but the performance is also better than the other two systems [6].

8.3.2 Practical Application: Internet of Underwater Things

Underwater Internet of Things (UW-IoUT) can be used for underwater environmental monitoring, navigation, pipeline monitoring, and other applications. Underwater MI communication technology can be transmitted within a range of several hundred meters, and the data rate is as high as tens of Gbps [159]. Therefore, it is widely used in IoT communication systems.

This chapter describes a deep-sea monitoring network in which underwater relays use MI communication. Due to remote communication, MI communication-based systems need relay nodes, which can effectively forward data. On this basis, in order to improve network throughput, an optimal relay placement scheme based on MI communication is proposed. A global optimization technique based on reverse convex programming (RCP) is used to carry out the non-convex optimization problem of relay placement. This method follows the concept of queuing network, and the goal is to use the least number of relays to obtain the maximum throughput of the underwater network [159].

In the following content, the system model, the use of global optimization technology to solve the placement problem, and the effectiveness of the method will be given.

Consider an IoT network based on MI communication for subsea monitoring. The model consists of relay nodes and convergence points, as shown in Fig. 8.14. Here it is assumed that the relay node is in a straight line and can obtain data from nearby nodes in a multihop manner. The relay node transmits the data to the receiving node and then to the ground base station, and the receiving node and the ground base station can pass through optical fiber for communication.

Since the transmission capacity of MI communication largely depends on the distance between the receiver and the transmitter, the SNR of MI communication is a function of the separation distance r. That is to say, the communication links between

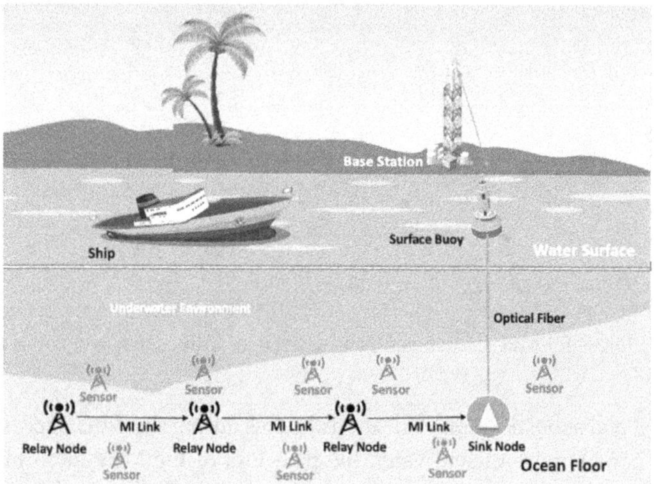

Figure 8.14 Proposed MI-based relay node placement for IoUT [159]

the relay nodes closer to the receiving node are responsible for placing a large amount of data to the upstream node, so these links require larger channel capacity [159].

Use a K-dimensional vector, where $K = (1,2,\ldots)$ with $y \in \mathscr{R}^K$. Consider that $M = 1,2,\ldots,M$ is the set of relay nodes on the subset of $L = [0,L]$. At the same time, it is assumed that the sink node is located at $X = 0$, denoted as X_0. And considering that the nodes on the seafloor are straight lines, it can be modeled as a one-dimensional area L near M relay nodes, for example, where $(m = 0,1,2,\ldots)$ represents the area covered by the mth node. According to a series of assumptions and derivations, the problem can be simplified to the following formula:

$$\max_{p\in\Re, r\in\Re^M} P$$

$$s.t. R(r_i) - \frac{pr_i}{2} - \sum_{m=i+1}^{M} pr_m >= 0, i \in 1,2,3,\ldots,M \tag{8.17}$$

$$p >= 0, \sum_{i=1}^{M} r_i = L, r_i >= 0, i \in 1,2,3,\ldots,M.$$

It can be noticed that the above formula does not converge to the feasible region, but the convex shape of the inequality is less than 0. Therefore, the convergent feasible region is indeed a supplement to the convex set. These constraints are also called reverse convex constraints [159]. RCP can be used to solve it, and the optimization problem can be changed to the following formula:

$$\max_{y\in\Re^K} u(y) f_i(y) >= 0, (i = 1,2,\ldots,M). \tag{8.18}$$

In order to find the optimal solution, the problem can be divided into subproblems, where the value of P is regarded as a non-variable, and then according to the local optimal solution expression of the subproblem, it is extended to the global optimal solution expression.

Table 8.2 Parameters used for simulations

Parameter	Value
Permeability μ	$4\pi \times 10^{-7}$H/m
Ocean water conductivity σ	4S/m
Transmitter power P_t	0.5 W
Noise power P_n	2×10^{-6} W
Operating frequency f	500 Hz
Area length L	500 m
Bandwidth W	5×10^3 Hz
Transmitter diameter D_t	0.2 m
Receiver diameter D_r	0.2 m
Number of relay nodes M	5, 10, 20
Number of coil turns N	100, 250, 500

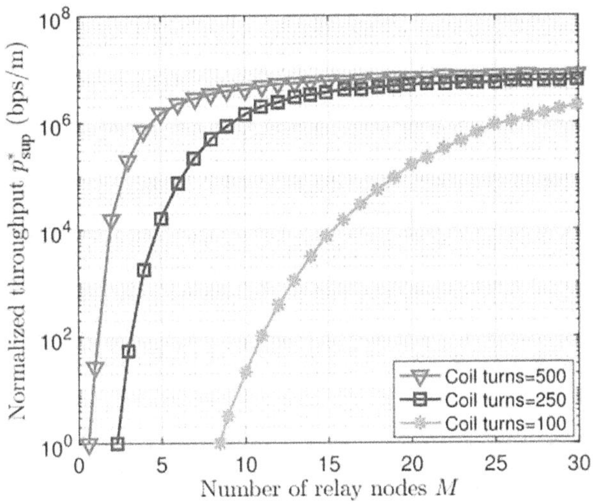

Figure 8.15 Impact of M number of relay nodes on the normalized throughput p_{sup} for optimal placement (L = 500 m) [159]

After simulation analysis, the feasibility of the optimal relay placement solution in the IoT network based on MI communication is verified. Due to the harsh underwater environment, it is necessary to find the optimal placement of relay nodes in order to increase the throughput of the network. Therefore, a reliable and economical underwater communication system has been developed that can transfer the acquired data to the surface for data analysis. The data parameters used in the simulation are shown in Table 8.2.

The effect of the number of coils and the number of relays on performance in an underwater environment is shown in Fig. 8.15. Figure 8.15 shows three different types of coil turns and the upper limit of the maximum normalized throughput of M relay nodes.

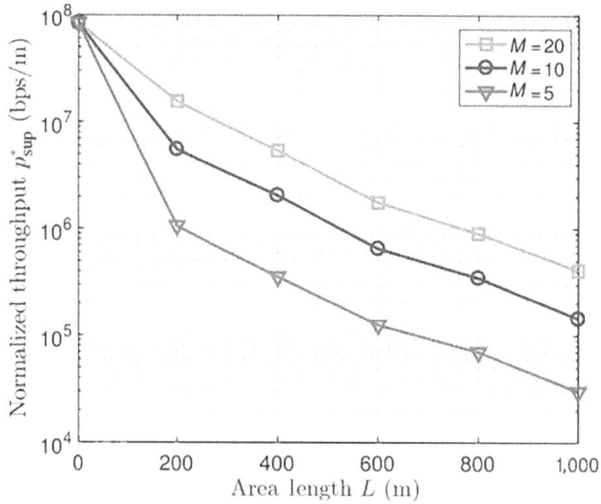

Figure 8.16 Impact of area length L on the normalized throughput p_{sup} for relay node with coil turns = 500 [159]

It can be seen from Fig. 8.15 that as the number of relays increases, standardized throughput is improved, and systems based on MI communication also depend on the type of coil and the number of turns of the coil. Increasing the number of turns of the coil can improve the normalized throughput. In addition, we can see from the figure that when the number of relays increases to a certain value, the standardized throughput reaches saturation. So, it can be concluded that as the number of turns of the coil and the number of relays increase, the standardized throughput will increase to the optimal value [159].

In Fig. 8.16, we calculated the maximum standard throughput of different numbers of relay nodes as a function of the area length L. From the figure, we can see that increasing the area length L reduces the normalized throughput, and when the L value is small, the standard throughput of different M values has a small difference, and the divergence between them increases with the increase of L [159].

Analyze the influence of the increase in the area length on the number of relay nodes when the communication range is 50 m. It can be seen from Fig. 8.17 that when L increases, more relay nodes are required in order to maintain the communication range unchanged.

In Fig. 8.18, the location of the relay node M according to the optimal deployment strategy is shown. This figure provides a graphical overview of the number of relay nodes as a function of the distance from the receiving node in a given area of 500 m, and the distance between relay nodes increases as the distance from the receiving node increases.

In this chapter, an optimal placement strategy for repeaters based on MI communication is proposed, which uses network queuing technology to model the system and optimizes placement to improve throughput [159]. This problem is an RCP problem, and an effective method has been developed to solve the global optimal solution.

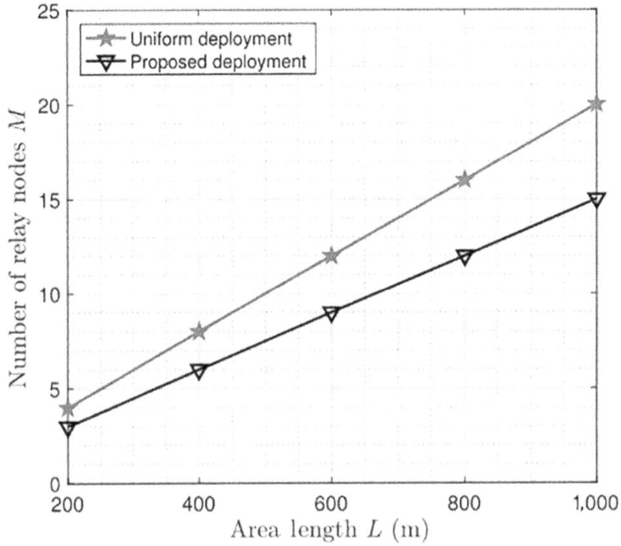

Figure 8.17 Number of relay nodes vs. area length with communication range 50 m [159]

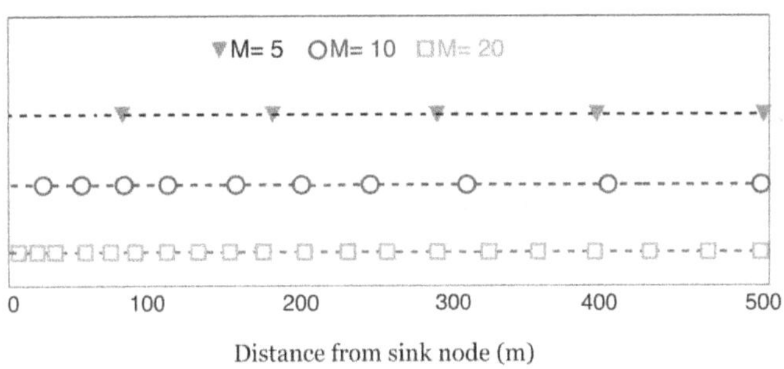

Figure 8.18 An illustration of the proposed optimum relay placement strategy with each point representing a single relay node in a maximum area length of $L = 500$ m [159]

8.3.3 AUV Based on MI

Utilizing the mobile characteristics of AUVs, we can transmit the information between underwater devices and realize the application of the UW-IoT. However, conventional acoustic communication has the disadvantages of high sensor cost and high error rate, so it is difficult to meet the needs of the UW-IoT. Wireless communication based on MI can overcome the above shortcomings. MI communication only relies on the near-field component of the magnetic field to transmit information, and the energy of MI communication is the resonance between the coil inductance and the resonant capacitor, so high power efficiency can be achieved using MI communication [160].

In addition to the above, the use of MI for communication also has the advantages of non-interference, security, and multipath. Due to the moving characteristics of the AUV and the influence of water flow, especially when the AUV is close to the target

Table 8.3 System parameters

Parameter	Value	Description
L_1, L_2	400 μH	Self-inductance of the coil antenna
C_1, C_2	$6nF$	Resonant capacitor
r_1, r_2	4 Ω	Intrinsic resistor of the coil antenna @100 kHz
U	1 V	Input voltage

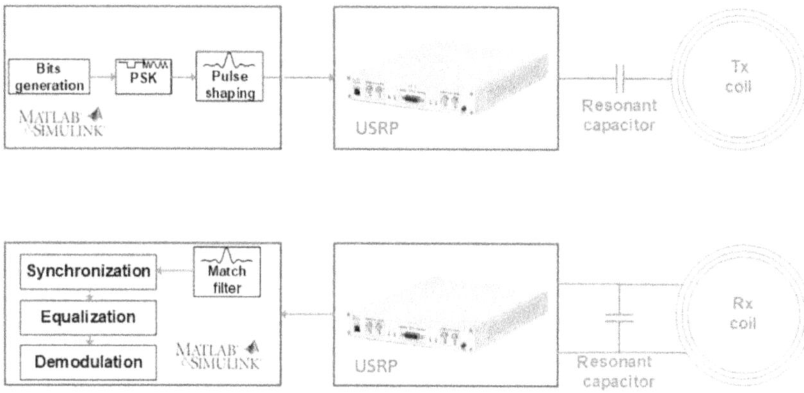

Figure 8.19 MI transceiver block diagram [160]

sensor, the magnetic field decays rapidly, the received MI signal introduces severe fluctuations, and the MI communication channel no longer remains static. In this chapter, a dynamic MI wireless communication system was established, and the experimental results were evaluated.

The designed MI system can work under different coupling conditions and make full use of the dynamic bandwidth. We have evaluated the performance through MATLAB/Simulink and then measured underwater through an MI test platform based on the universal software radio peripheral (USRP) system's performance in a 200-gal tank.

The first is the design of the MI transceiver: The block diagram of the proposed MI transceiver is shown in Fig. 8.19, and the parameters of the coil antenna and resonant capacitor are shown in Table 8.3. The carrier frequency is set to 102.8 kHz. Since MI communication is a narrowband communication technology, phase shift keying (PSK) and quadrature amplitude modulation (QAM) are suitable modulation schemes for MI communication. Then use MATLAB for signal processing, and USRP is responsible for the conversion from analog to digital and vice versa.

The overall simulation model is shown in Fig. 8.20. At the same time, for better experimental results, we divide the system into three parts, namely the transmitter part, the receiver part, and the MI channel part.

In the simulation process of the MI channel, use mutual inductance to simulate the coupling between the two coils and use the control voltage source to couple the modulation signal to the resonant circuit. At the same time, the free load receiver design is adopted to obtain the highest SNR. Since our purpose is to verify the available

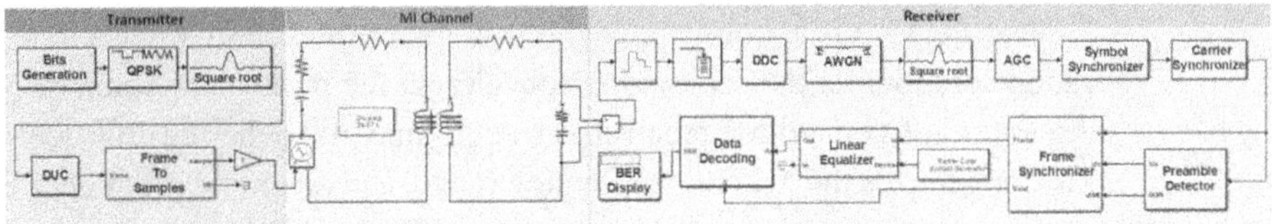

Figure 8.20 Simulation model of the MI wireless communication system [160]

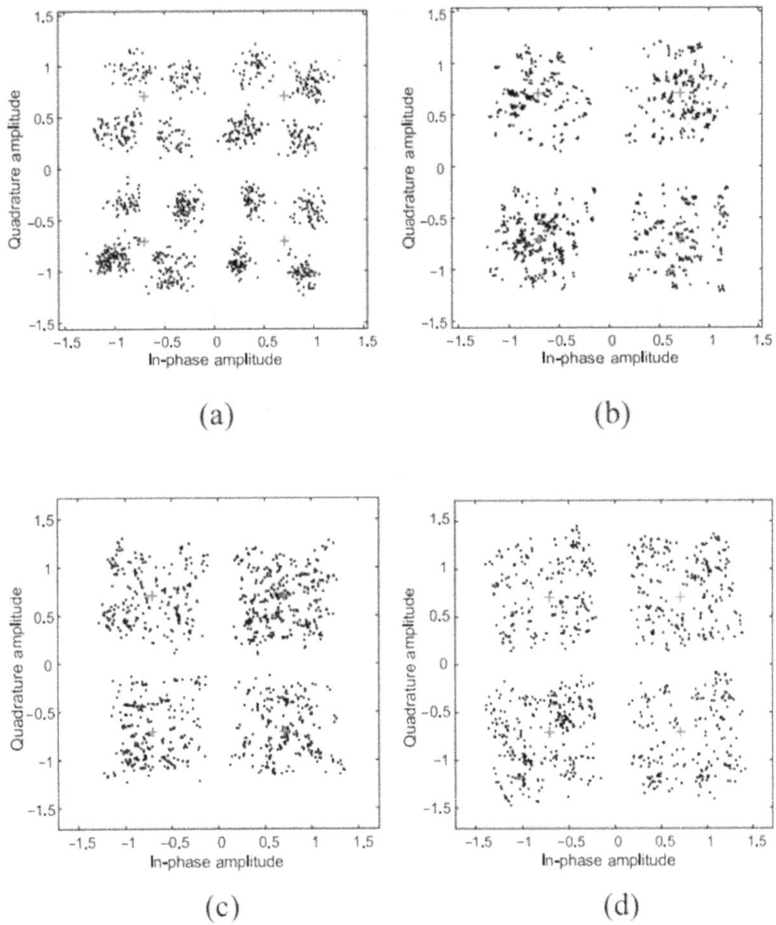

Figure 8.21 Mobility impacts to the demodulation results of the MI wireless communication system with the maximum achievable symbol rate and zero BER under different coupling coefficients. (**a**) Demodulation results @ $k = 0.0001$ with the symbol rate of 3,200 sps. (**b**) Demodulation results @ $k = 0.04$ with the symbol rate of 8,000 sps. (**c**) Demodulation results @ $k = 0.06$ with the symbol rate of 10,667 sps. (**d**) Demodulation results @ $k = 0.08$ with the symbol rate of 8,000 sps [160]

bandwidth of the MI wireless communication system under different coupling conditions, no equalizer is applied at the receiving end in the simulation. Figure 8.21 shows the constellation diagram output by the carrier synchronizer when the maximum symbol rate is reached with zero error rate under different coupling conditions. The simulation analysis and prediction results are consistent.

Near-field MI is a promising low-power underwater short-wave wireless communication solution. At the same time, considering the rapid attenuation of MI signals in the water, a free-load MI receiver is used to maximize the sensitivity of the receiver. This maximizes the MI communication range, and due to the frequency division phenomenon, the available bandwidth of the MI channel is also very sensitive to the coupling conditions between the MI transmitter and the receiver [160]. Finally, the proposed dynamic MI bandwidth is verified by simulation.

8.3.4 Conclusion

In summary, underwater MI communication has the following advantages compared with traditional acoustic communication:

(1) Small signal propagation delay: MI waves propagate fast in water, which can significantly improve the delay performance of underwater communication and help facilitate the realization of underwater network protocols.
(2) Predictable and stable channel response: The high-speed propagation of MI waves can weaken the frequency offset caused by the Doppler effect. At the same time, the transmission range of MI communication and the quality of the channel are not affected by other factors such as water pollution.
(3) Large transmission range and high data rate: A coil of any size on the MI transmitter can generate a time-varying magnetic field, and the MI receiver can also capture low-frequency signals. In addition, the working frequency of the MI coil can reach megahertz bands, which makes the transmission data rate high.
(4) Stealth operation: Use non-sound and invisible MI waves for communication.
(5) Collaborative sensing and tracking with underwater swarming robots: Underwater robots can coordinately track the source of pollution or pests with faster convergence speed and accuracy.
(6) Real-time underwater surveillance: A large amount of multimedia content and data can be transmitted at high speed.
(7) Disaster assessment and rescue in a chaotic aquatic environment: In a chaotic underwater environment, traditional acoustic communication will experience severe multipath fading, resulting in a serious degradation of channel quality [158].

8.4 MI-Based Underground Utility-Line Monitoring

As the safest, fastest, and cheapest way to transport a large amount of liquid over a long distance, the underground utility line is widely used all over the world to transport fresh water, crude oil, and natural gas. It is regarded as the lifeline of the national economy in many places. However, when the underground pipeline is exposed to a variety of adverse soil environment for a long time, it is very vulnerable to soil environment corrosion or man-made damage, resulting in pipeline leakage. Pipeline leakage will not only cause great economic losses but also cause serious pollution to the surrounding environment. Therefore, an easy, deployable, flexible, effective, and inexpensive approach for real-time distributed data acquisition and monitoring is extremely important for underground utility-line management. Managers can effectively detect the underground utility-line in real time through underground utility line monitoring, so as to minimize losses.

In underground wireless communication, the attenuation of EM wave is much greater than that of the magnetic field. In order to ensure the connectivity of the network, high-density underground sensors are needed, which may lead to high deployment and maintenance costs. In the case of mining accident or explosion, some underground roads are cut off, and the traditional wired communication and wireless communication cannot work normally. Communication based on MI is a promising signal propagation technology in the soil medium, because a dense soil medium will not cause a higher magnetic field attenuation rate than in air, and MI channel conditions will not change dramatically with the change of soil properties. In addition, since the magnetic field frequency used is low and the wavelength is about a kilometer, there is no multipath effect.

In recent years, MI-based underground utility-line monitoring has developed rapidly. The communication based on MI, combined with the sensor network composed of high-precision pressure sensors and humidity sensors and technologies, such as underground to above ground radio distribution, can realize the real-time detection of underground utility-line monitoring and accurately locate the leakage of underground lines according to the detected signals. Relying on its high detection accuracy, reliable communication performance and low application cost, MI-based underground utility-line monitoring is gradually becoming the mainstream of underground utility-line monitoring.

8.4.1 Existing Underground Utility-Line Monitoring Techniques

The term leakage is defined as a way of fluid to escape outside pipes, tanks, or containers [161]. The leakage of underground utility line will cause a lot of energy or fresh water to be discharged into the soil, waste resources, and bring great economic losses. If the leaked liquid is toxic or corrosive, the leakage will also cause great disasters to the environment. In the actual underground utility-line operation, the risk of pipeline leakage always exists. Old and weak pipeline structures, poor valve maintenance, and insufficient corrosion protection are all factors leading to leakage. Changes and effects of external conditions, such as ambient temperature, pipeline pressure, and traffic vibration will lead to pipeline leakage. With the continuous expansion of the underground utility line network, the possibility of leakage increases, and the demand for leakage monitoring becomes more and more unrealistic.

The traditional underground utility-line monitoring mainly depends on the regular inspection of maintenance workers, which is inefficient, and the labor cost is very high. When the underground utility-line network is large, this working mode cannot monitor the pipeline in real time, nor can it locate the leakage location quickly and accurately. Once leakage occurs, it will lead to greater economic losses and environmental pollution.

At present, the rapidly developing high-precision sensors can detect the subtle changes in the environment of the underground utility line network in real time and transmit them to the adjacent data-processing center for analysis, processing, and leakage warning through wireless communication technology. This process is fully automated to facilitate the timely detection of leaks. The existing underground utility-line monitoring technologies have two types: The sensor is placed inside the pipeline and the sensor is placed outside the pipeline according to the location of the sensor.

(1) Ground Penetrating Radar

A ground penetrating radar is a special radar especially designed to detect the condition of underground pipelines without diggings. The attached range can reach several miles. It is generally inherited from portable equipment, which is convenient for maintenance workers to carry and use. However, the ground penetrating radar cannot really monitor the pipeline leakage in real time and still needs intensive labor input [162].

(2) Soil Properties Sensors

Once the pipeline leaks, the fluid in the pipeline overflows into the soil, which will inevitably lead to changes in the characteristics of the soil near the pipeline. Soil sensors can monitor the changes of soil properties and transmit the signal to the data-processing center through the wireless communication system to quickly identify the abnormalities of soil properties and accurately locate the location of pipeline leakage. The type of sensor used is determined by the transmission fluid of the monitored pipeline. The temperature sensor can be used to detect the leakage of high temperature fluid such as hot water. The pH sensor can be used to detect the leakage of strong acidic or strong alkaline chemical products. A soil dielectric property sensor can be used to detect the leakage of the crude oil pipeline. The development of sensor-based monitoring technologies is mainly limited by the development of its communication system. In the underground environment, the cost of laying the wired communication channel is too high, and it is easy to cause communication interruption due to landslides and other reasons. In addition, the signal attenuation of EM wave communication in an underground environment is too large, and there is obvious multipath effect, so the reliability of the communication system is not high. Therefore, MI-based underground utility-line monitoring is the mainstream trend of future development.

(3) Acoustic Devices

Acoustic devices are applied to underground utility-line monitoring, which is mainly used to monitor the high-frequency oscillation in the pipe wall caused by small leakage in the pipe. By tracking the source of vibration data, the location of pipeline leakage can be accurately located [162]. Due to the limitation of detection range, it is usually necessary to install high-density acoustic sensors in the pipeline to cover the whole pipe network, which leads to huge difficulty in maintenance. In addition, acoustic sensors are not sensitive to large leakage, because they will not vibrate in the high-frequency characteristic frequency band, so acoustic devices are generally only used to detect small leakage on underground pipelines near inspection stations or pump stations. Generally speaking, too dense deployment of sensor nodes in the pipeline will increase the risk of pipeline leakage, so the deployment location of sensors in the pipeline is also limited to the pipeline of an inspection station or a pump station.

(4) Transient-Based Methods

Transient-based methods is a popular transient pipeline detection method, which can be divided into the following four steps [162]:

(1) For some special reasons, the pipeline leaks in an instant.

(2) Pressure sensors deployed at pipe network checkpoints measure pressure during transient events.

(3) The measurement data are transmitted to the data-processing center through the communication system.

(4) According to the measured data, a transient simulation model of underground utility line is established.

This method can quickly and accurately analyze whether the pipeline leaks and the specific location of the leak. According to the data detected by the high-precision pressure sensor, it can also predict the scale of the leak according to the corresponding mathematical model and give a reasonable emergency plan. However, because the pressure sensor can only be deployed in a limited location, the transient-based method cannot provide sufficient leakage detection and positioning accuracy.

8.4.2 MI-Based Underground Utility-Line Monitoring

(1) System Architecture and Operational Framework

The system is a clustered architecture, and it consists of heterogeneous sensors. The cluster is formed in each pipeline section [161]. As is shown in Fig. 8.22, the sensor network in the system architecture of MI-based underground utility line monitoring is mainly composed of the hub layer and the in-soil sensor layer, which are deployed in different locations to jointly detect whether the pipeline is leaking.

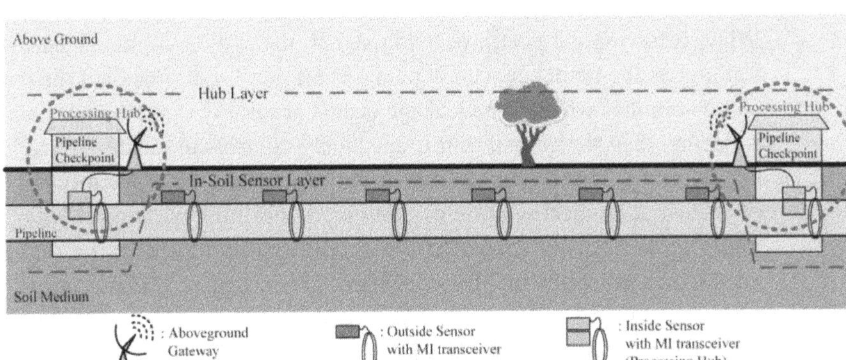

Figure 8.22 System architecture of MI-based underground utility-line monitoring [162]

The hub layer consists of pressure sensors and acoustic sensors deployed inside the pipeline. The in-soil sensor layer consists of sensors deployed outside the pipeline to measure soil properties, such as the humidity sensors and the temperature sensors. All sensors require high precision and low power consumption. The detected data are transmitted to the head cluster at the checkpoint or pump station, and then the head cluster through communication based on MI. Then, the head cluster transmits the data to the remote management center for data processing, analysis, and decision-making.

(2) Wireless Sensor Networks

In the hub layer, the sensors used are mainly pressure sensors and acoustic sensors that are deployed in the pipeline of the checkpoint and pump station. The pressure sensor uses a transient-based method to identify areas where there may be leakage in the pipeline. The acoustic sensor is used to monitor the high-frequency oscillation caused by small leakage in the pipe wall [162]. The detection advantages of the two sensors have their own emphasis and complement each other.

In the in-soil sensor layer, the sensors used are mainly humidity sensors and temperature sensors that are deployed in the soil outside the pipeline along the laid pipeline to detect the changes of soil properties. Increasing the sensor density can effectively improve the detection quality and solve the problems of small detection range and low accuracy, but it will also greatly increase the maintenance cost.

The above internal sensors form a sensor network, and the two measurement methods complement each other. There are abundant pressure sensors in the inspection station and pump station, which can continuously monitor the pipeline. Excessive deployment of pressure sensors inside the pipeline will also increase the risk of pipeline leakage. Therefore, in the case of low deployment density, the detection accuracy and positioning accuracy of pressure sensors are insufficient. The deployment density of soil sensors is relatively large, and the detection and positioning effect is relatively accurate. However, due to the limitation of battery power supply, the gap working mode is generally adopted, which cannot be monitored continuously.

High-precision sensors in sensor networks are equipped with MI transceivers. Any pipe rupture or leakage can be detected by monitoring the pressure of the fluid flowing in the pipe. The sensor transmits the data acquired by the pipeline to the ground through a wireless connection, which means that it does not need any physical connection with the ground. There are stations on the ground to receive and process signals. When receiving the post-processing signal, the station communicates with the central unit as the main connection point. The central unit processes the received signal and determines whether the leakage occurs according to the corresponding model. If it occurs, then the central unit immediately controls the underground utility line to stop working to reduce economic losses. Compared with the current working mode of industrial application, this working mode has high accuracy, high efficiency, and high degree of automation. The pressure sensor and the acoustic sensor in the hub layer communicate with the soil characteristic sensor buried outside the pipeline by means of MC. The soil characteristic sensor cooperates to process the measured data and wirelessly transmits the data to the processing center located at the checkpoint and pump station along the underground pipeline in a multihop manner.

Because the inspection points of underground pipelines and pump stations usually have aboveground parts for maintenance, the internal sensors of each inspection point or pump station are connected with the aboveground gateway wirelessly or through a wire and then send the measurement data to the remote management center in a multihop mode or through satellite communication.

In order to more accurately detect and locate the location of possible leakage and realize long system operation life, a three-phase collaborative detection strategy is generally adopted for analysis [162]:

(A) The pressure sensor measures the flow pressure during the transient event through the transient-based leakage detection method and sends the measurement results to the remote management center. Through the calibration of the pipe network transient simulation model based on pressure measurement and combined with deep learning technology, the suspicious areas with possible leakage are identified.

(B) The management center notifies the pressure sensor of the pipeline in the suspicious area. The pressure sensor sends a data request to the soil characteristic sensor along the pipeline. These soil characteristic sensors remain in sleep mode to save battery energy until a data request is received. The soil characteristic sensor is activated from the sleep mode according to the data request, measures the required soil characteristics, and then sends the measured data to the pressure sensor by means of MI technology in a multi-hop manner.

(C) After receiving the soil property measurement data along each suspicious pipeline, the adjacent processing centers exchange these high-resolution measurement data through the ground wireless channel. The adjacent processing hub determines whether there is leakage in the suspicious pipeline through processing cooperation in the network and locates the leakage location.

(3) Magnetic Induction-Based Underground Communication

Since the underground wired communication line is easily interrupted by factors such as traffic vibration or collapse, more reliable and efficient wireless communication is generally used between the soil characteristic sensor and the processing center. In the underground communication environment, the traditional EM wave technology attenuates too much because it is absorbed by soil, rock, and groundwater. The soil, groundwater, and most rocks have relative permeability to the magnetic field, and the signal attenuation is very small. MI attachment is directly proportional to $1/R$, while EM wave's attachment is directly proportional to $1/R^3$, where R represents the communication distance. In addition, the MI channel does not change with the change of soil properties, and there is no multipath effect. Therefore, using MI communication is more reliable and cheaper.

For the magnetic channel, frequency-dependent absorption (skin effect) and background noise need to be considered, and they jointly determine the working frequency of MC [163].

The surface effect refers to the frequency-dependent effect of the conductivity of rocks and other objects on the magnetic field. This is because the time-varying magnetic field will produce an electric current in the rock, and the function of the electric current is to counteract the magnetic field. Therefore, even if the permeability

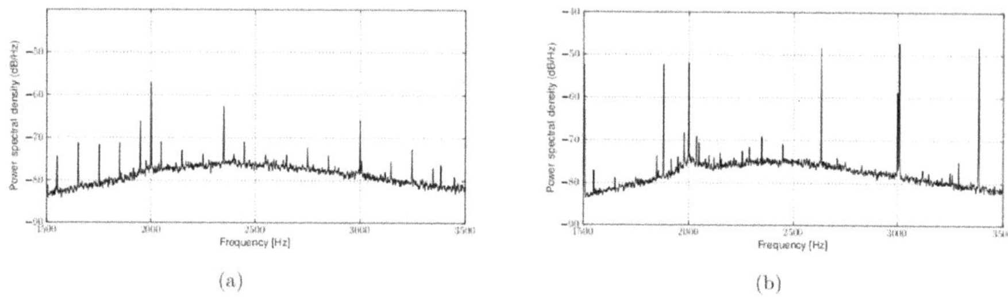

Figure 8.23 PSD of background noise. (a) The PSD as captured by the x-axis sensor. (b) The PSD as captured by the z-axis sensor [163]

of the medium is close to 1, the skin effect begins to dominate the channel over a long distance. The skin depth, i.e., the distance at which the signal attenuates to $1/E = 0.368$, is

$$\delta = \sqrt{\frac{2}{\omega\mu_0\sigma}}, \tag{8.19}$$

where ω represents the frequency, $\mu_0 = 4\pi \times 10^{-7}$H/m represents free-space permeability, and σ represents the medium conductivity. Equation (8.19) shows that the higher the carrier frequency, the greater the attenuation of MC.

Background noise refers to low-frequency noise caused by various natural or man-made reasons, such as electrical noise and ground resonance. Noise can be approximated as roughly following a $1/f$ distribution, with strong narrowband interferers. The PSD of background noise in the range of 1,500–3,000 Hz is shown in Fig. 8.23 for the x and z channels. Also, note that since the noise source is directional, the harmonics of x and z channels are different. In order to avoid strong interference, the channel can be concentrated between adjacent harmonics [163]. At lower MC frequency, the wavelength is much larger than the distance between nodes. Subcommunication is carried out in a reactive (quasi-static) near field. The field is non-propagating and hence there can be no multipath or fading.

In an MI communication system, relay points are usually added between the transmitter and the receiver to expand the communication range of MI in the underground environment. MI relay point generally adopts simple coil without any energy or processing device. The specific structure is shown in Fig. 8.24. In the first relay, a sinusoidal current is introduced into the transmitter coil, and then a sinusoidal current is induced in the second relay. If this cycle continues, these relay coils form MI waveguides in the underground environment to guide the so-called MI wave as waveguides to realize the communication between relays. It should be noted that if the pipe is made of metal, no (or very few) relay coil is required, because the metal pipe itself is the magnetic core of the MI waveguide. Relay coils are required only when the pipe is nonmetallic. It only needs to deploy one relay coil, about 5 m away from each other. The underground pipe is the perfect core for winding these coils. If the coil is installed on the pipe during deployment, the cost of deploying the coil is small.

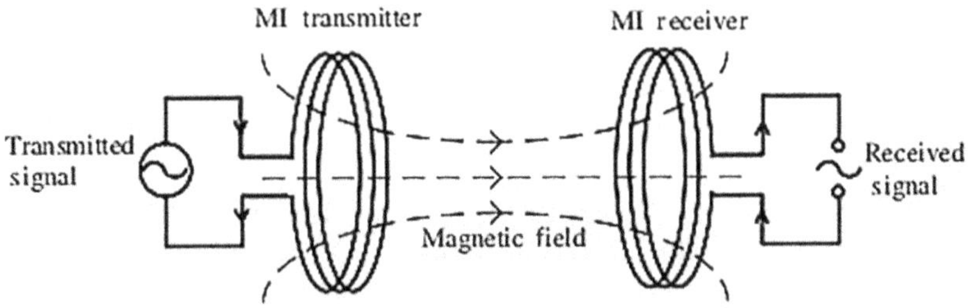

Figure 8.24 MI-based communication [162]

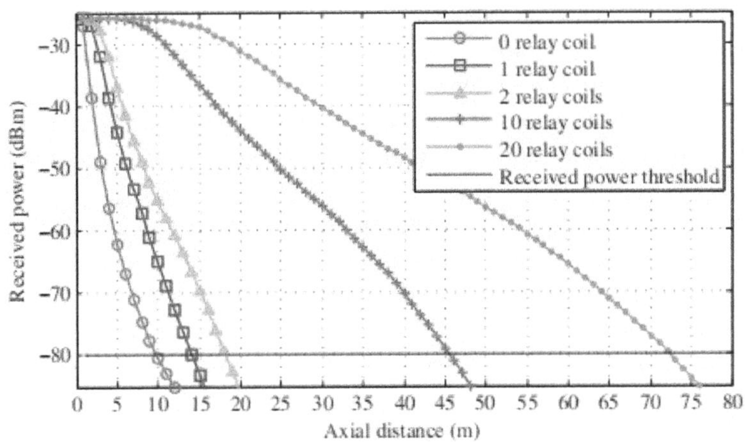

Figure 8.25 Received power of MI waveguides with different numbers of relay coils [162]

In order to save the deployment cost, the optimal number of relay coils between the two sensors can be derived according to the required bandwidth and the distance between the sensors. In Fig. 8.25, the received power of the 10 MHz + 0.5 kHz signal using MI waveguides with different numbers of relay coils is shown as a function of the link length *d* [163]. The axial communication range of an MI waveguide with a certain number of relay coils is shown as the intersection point of the received power and the 80 dBm threshold. The axial communication range increases with the increase in the number of relay coils. However, as the number of relay coils increases, the increase in communication range caused by additional playback coils decreases.

The MI waveguide has the following four advantages in underground communication of mise-pipe:

(A) By using MI waveguides, the number of sensors required may be significantly reduced. Specifically, by properly designing the waveguide parameters, the total path loss can be greatly reduced.

(B) The deployment and maintenance of the MI waveguide are more convenient, and the system stability and reliability are good. Unlike sensor equipments, once the relay coil is buried underground, no additional maintenance is required. Even if some coils are damaged in extreme cases; the remaining coils still provide reliable network operation.

(C) The relay coil does not consume energy, and the unit cost is very small. Therefore, an MI waveguide is an ideal choice for an underground pipeline monitoring system.

(D) Because the MI waveguide can be used to inductively charge the downhole sensor, the service life of the system can be greatly prolonged.

(4) Wireless Signal Networks

The above MI-based WSN can check the leakage through the value measured by the sensor. However, it can only provide the point measurement value and then send the point measurement value and compare it with the reference value. In addition, these reference values need to be calibrated according to possible changes in the environment to ensure correct detection. The concept of wireless signal network (WSiN) appears as an improved technology, which provides a more accurate and reliable improvement for underground utility line monitoring.

WSiN to detect pipeline leakage is realized by detecting the strength of the wireless signal. According to Eq. (8.19), when the properties of soil change, such as soil water content, the communication channel will be changed, thus affecting the wireless signal. The RSS information can be used as an index of soil physical change, so it can be used to monitor the leakage in the sensing area in real time.

The WSiN nodes also use batteries for power supply, so they also use sleep mode and low-power design to reduce power consumption. In addition, since the attenuation rate is very high in the soil, the communication radius will be very small and dependent on the operating frequency. Three methods can be adopted to increase the communication radius: Using high transmission power, high gain antenna or working at low-frequency, but considering the effective power consumption, low frequency operation is the only feasible solution. When the transmission power is less than 1 MW, the communication radius in soil can be increased from 20 cm to 25 m when working at 2.4 GHz and 160 kHz [161]. The change of soil water content indicates that leakage may occur under normal conditions. In order to evaluate the effect of soil water content change on RSS, the RSS data under different water content were collected in this experiment. The experimental results show that the dielectric constant and conductivity increase with the increase of water content, which will lead to more attenuation of received signal strength. When the transmission power is less than 1 MW, the communication radius in soil can be increased from 20 cm to 25 m at 2.4 GHz. Experiments show that the dielectric constant and conductivity increase with the increase of water content, which will lead to more attenuation of received signal strength [161].

8.4.3 Future Development

The scale of the underground utility-line network system is becoming larger and larger. Therefore, it is necessary to optimize the deployment of sensors, comprehensively consider the hardware cost and measurement effect, flexibly select different types of sensors according to the specific environment, and determine the number and location of different devices with an optimal deployment strategy, so as to minimize the deployment and operation cost, and maximize leakage detection accuracy.

In addition, the adaptive equalization of the MI transceiver should be done well. When the coil wire resistance is small, the MI channel will experience frequency selective fading. In order to ensure the low path loss of the MI channel, the coil line resistance should not be too high. Therefore, the frequency selectivity of the MI channel is inevitable, so adaptive equalization should be done well. For the MI waveguide channel, several existing equalization strategies need to be studied, including linear equalization, decision feedback equalization, Tomlinson Harashima precoding, and grid-based equalization [162].

MI-based underground utility-line monitoring is also an important branch of underground utility-line monitoring in the future. MI-based communication technology greatly improves the reliability of the system and has been proved to be a research direction with great development potential. Of course, more research work is needed to effectively improve the communication radius and system performance.

8.5 Oil Reservoir Monitoring

The crude oil industry plays an import role in providing energy to support out daily life. The crude oil extraction is a challenging task since it is performed deep underground. As shown in Fig. 8.26, the oil extraction process can be divided into three phases. In the first phase, a wellbore is drilled into deep underground. Then, in the second phase, fracturing technologies are employed to open several fractures, so that crude oil can flow into the wellbore through the fractures. However, the fracture itself is not strong enough due to the harsh underground dynamics. To support the fracture, proppants will be injected during the fracturing process. In the last phase, crude oil is extracted from the wellbore to aboveground.

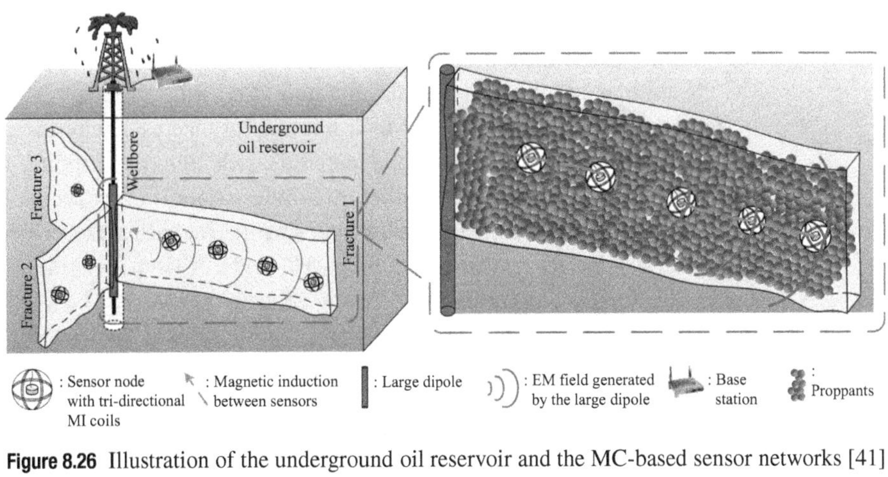

Figure 8.26 Illustration of the underground oil reservoir and the MC-based sensor networks [41]

As crude oil flows out of the oil reservoir, the pressure inside the fractures gradually decrease and crude oil no longer flows with natural pressure. To address this issue, at the other end of the fracture, water can be injected to increase the pressure and push crude oil into the wellbore. This process requires a significant amount of information, such as the pressure inside the fracture, mixture content in the oil reservoir, among others. These pieces of information can be effectively collected by using sensors. The sensing data can be collected at the aboveground control center to intelligently manage the oil extraction process.

Different from terrestrial environments, it is difficult to deploy sensors in the deep underground fractures. The fracture also has very narrow space; its width is approximately 1 cm. Also, the harsh environment does not allow human beings to perform any operation inside. Thus, we need autonomous approaches to deploy and manage these sensors.

First, wired sensing technologies are cumbersome for deployment. In such a harsh environment, we need to provide additional protection to avoid sensors and wires being destroyed by the salty water or crude oil. If there is any failed sensors or wires, the system is hard to repair. Once they are deployed, it is almost impossible to fix these issues.

Therefore, wireless solutions are more preferable since they are easy to maintain and require less labor to deploy. Different from wireless communication in the air, the mixtures of crude oil and water, along with the inhomogeneous lossy media, create significant challenges for wireless signal propagation. The RF signals experience significant attenuation losses, which results in an extremely short communication range. Also, the inhomogeneous media make the signal propagation hard to predict. The terrestrial wireless communication channel model fails in such kind of environment.

Beside RF signals, wireless communication can also be achieved using various other technologies, such as the acoustic signals. In [164], acoustic signals are used to communicate with deep underground sensors. However, the acoustic signal propagation relies on a steel wall and usually requires high transmission power. This requires infrastructure support, and the high power may cause heating issues in the oil reservoir. This can be a great threat to the whole system.

An MC-based wireless self-contained sensing system is proposed in [41]. As shown in Fig. 8.26, the sensing system consists of three parts. First, wireless sensors with MC capabilities are injected into the fracture along with proppants. The proppants are used to support the fracture, which has similar size as a wireless sensor. Also, these proppants are made of magnetic materials with high permeability. This can create a high-efficiency magnetic channel for wireless signals to reduce the propagation loss. Since wireless sensors are deployed during the fracturing process, no additional efforts are needed for sensor deployment. However, it is challenging to ensure that sensors are deployed properly. Without any control, sensors may be injected into random locations. This issue is addressed by tethering sensors that ensure that wireless sensors are deployed along a one-dimensional line.

Even if sensors are deployed in fractures, there are still two challenges to design self-contained sensor networks. First, the sensing data need to be sent to the aboveground using MC technologies. The sensors form a multihop relay network and send data to the wellbore. Then, the wellbore is connected to the aboveground control center that can monitor and control the underground oil extraction process in real time.

Those sensors in fractures cannot be maintained using traditional technologies since they are not accessible once deployed. Thus, energy harvesting is a promising solution. An antenna is placed in the wellbore to radiate EM fields into fractures, and wireless sensors harvest energy from fields. Once sensors are charged, they will start sending data back to the wellbore. Due to the tiny sensor size, the sensor coils too are small, and hence are not efficient in energy harvesting. Thus, it takes a long time to charge before sensors can send data back. Implementations of the wireless sensors that can be put into an oil reservoir have been developed in [165, 166].

There are a few open research problems for the practical applications. First, the sensor size needs to be reduced further. The fracture is very narrow that cannot allow large sensors. Second, the tethered wireless sensors may not be placed at the designed locations. In such a case, we need to inject more sensors to ensure that the targeting areas are covered. With more sensors, random locations for a 2-D multihop network need to be considered.

8.6 Wireless Power Transfer

Recently, near-field wireless power transfer (WPT) has been widely used in the mobile device and sensor nodes charging in communication networks. Magnetic resonant coupling WPT is one of the most popular configurations that utilizes the mutual coupling of TX and RX to deliver power. However, when the charging distance increases, or the position of the RX is not aligned, the coupling effect is weakened, which causes low power transfer efficiency (PTE).

Thus, maximizing the achievable PTE in dynamic WPT scenarios becomes one of the key issues in the WPT systems design. In [167], it proves that a PTE upper bound of 50 % can be achieved by conjugate impedance matching on the TX and RX sides. Work done in [168] shows that a theoretically 100% PTE can be realized through optimal load design when the charging distance is close enough. However, these works do not address the problem of low PTE brought by the variations of charging position and distance. A magnetic MIMO technique is proposed in [42], where the amplitude and phase of the source currents are manipulated to achieve magnetic beamforming for a MIMO WPT system; the experimental results show that the overall charging time is shortened by increasing the PTE. These magnetic MIMO schemes give the optimal WPT solution when the RX is on a particular position, but the systems need accurate magnetic channel estimation and control of current in the charging source, which cannot be realized when the charging source is unchangeable or cannot be manipulated in extreme environments. Besides, the long-distance PTE is still low due to the significant path loss. Passive metasurface-enhanced WPT in [169] shows that a high PTE is obtained within a 2-m range in a point-to-point power transfer case. However, the directions and positions of power TX and RX are constrained strictly. Nowadays,

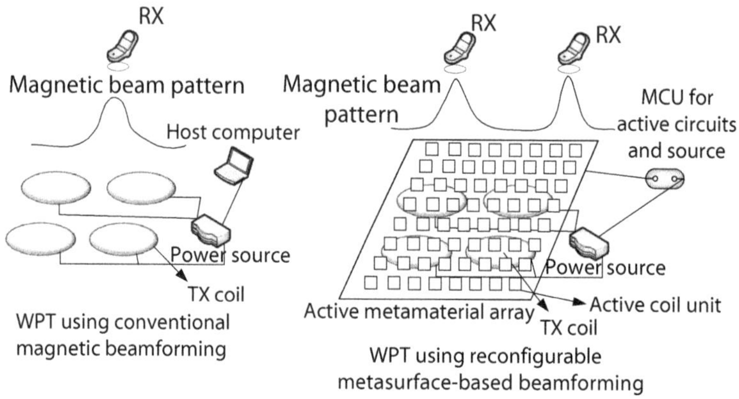

Figure 8.27 System architecture of wireless power transfer using reconfigurable metasurface-based beamforming vs. conventional magnetic beamforming [170]

in those extreme WPT scenarios, none of the existing WPT schemes can satisfy all the dynamic requirements: (1) Maximum PTE is desired. (2) The configuration of the power source cannot be manipulated. (3) The host cannot communicate with the RX to acquire the channel information.

In this subsection, we introduce an MI beamforming scheme using a reconfigurable metasurface, which can improve the PTE of the MIMO WPT system in dynamic scenarios. A detailed version of this subsection with technical part is given in [170].

As shown in Fig. 8.27, we use a reconfigurable metasurface to complete the magnetic beamforming process instead of manipulating the TXs in conventional magnetic beamforming. The reconfigurable metasurface consists of an active metamaterial layer, which is settled above the power source, and the MCU controls the active circuits. Inspired by our work in [171], the non-foster active circuit is used to obtain negative resistance and capacitor to control each coil unit on the metamaterial layer. In order to achieve the magnetic beamforming, when the receivers are in power receiving mode, the MCU measures the current in each TX to process the magnetic channel estimation, while the voltage of each TX is fixed. After the channel estimation, the MCU knows the receivers' positions and runs the algorithm to derive the optimal circuit parameter configuration of each unit in the active metasurface. Then the magnetic beamforming is achieved through the MCU, varying the impedance profile of the metamaterial layer. Our reconfigurable metasurface-based beamforming scheme does not need to manipulate the power source, which makes our system workable in extreme WPT conditions. Besides, we only use the MCU instead of the host computer in the conventional beamforming system, which reduces the cost. As shown in the analytical and simulation results, the proposed MI beamforming scheme can achieve two times the PTE of the conventional magnetic beamforming scheme in a 40-cm charging distance.

The system architecture of the proposed reconfigurable metasurface-based magnetic beamforming system is shown in Fig. 8.27. The conventional magnetic

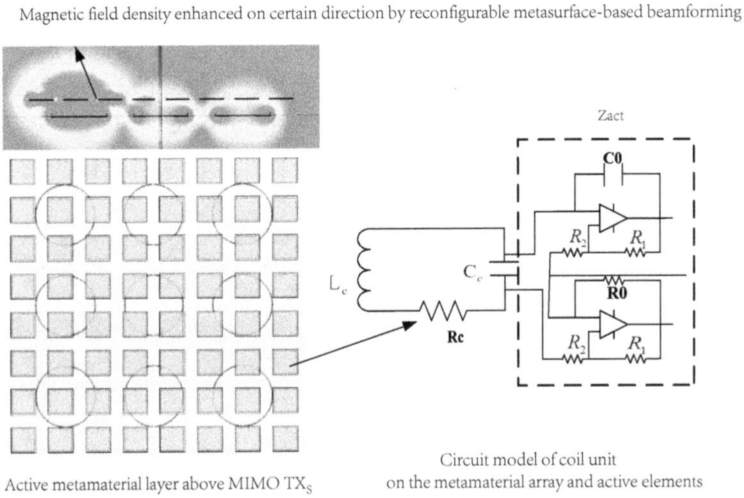

Figure 8.28 The topology of active metamaterial beamforming layer with the circuit model of coil unit on the array and active elements [170]

beamforming scheme for WPT requires the host computer to manipulate the current distribution in different TXs. Meanwhile, the RX needs to communicate with the host computer for magnetic channel estimation. The idea of our reconfigurable metasurface-based beamforming scheme is different from conventional magnetic beamforming. We are using an active metamaterial array that consists of coil units with active circuits. The metamaterial layer is put above the MIMO WPT TXs and connected to the MCU. The MCU does not need to control the current on each TX or communicate with RXs. Instead, the MCU is used to vary the parameter profiles in active circuits and measures the current in each TX for magnetic beamforming.

Metasurface is proved to be effective for improving the PTE in [169]. And work done in [172] gives a magnetic beam focusing solution in the subwavelength scale to concentrate the magnetic field in a particular direction. Our works in [135] and [173] prove that magnetic metamaterial can be manipulated through active circuits. And the dynamic magnetic gain can be obtained through active metamaterial. Inspired by those works, we design a reconfigurable metasurface-based magnetic beamforming system that can focus on the magnetic field generated by TXs in certain directions. The reconfigurable metasurface consists of an active metamaterial layer and the MCU. The metamaterial layer, which consists of a number of m identical square coils, is settled between the MIMO power TXs and the RXs, as shown in Fig. 8.28. The coil unit is a transfer model of the split resonant ring and is modeled as a resistance R_c, capacity C_c, and inductance L_c in series. Since the beamforming process requires varying the impedance value of each coil unit, the active circuits are introduced. One active impedance circuit consists of fixed resistance R_0, capacitor C_0, two adjustable resistances R_1 and R_2, and two ideal operation amplifiers. The total active impedance is $Z_{\text{act}} = -\frac{R_2}{R_1}R_0 + \frac{jR_1}{\omega R_2 C_0}$, which contains negative resistance and negative capacitance. Then the total impedance of the coil unit after applying the active element is $Z_c = R_c + j\omega L_c - \frac{j}{\omega C_c} + Z_{act}$, which can be controlled by varying the resistances R_1 and R_2 through the MCU. The magnetic beam can be steered by configuring different impedance values on each coil unit, as shown in Fig. 8.28; the magnetic field intensity is enhanced in a particular direction through beamforming.

References

[1] I. Akyildiz, D. Pompili, and T. Melodia, "Underwater acoustic sensor networks: Research challenges," *Ad Hoc Networks*, vol. 3, no. 3, pp. 257–279, 2005.

[2] L. Lanbo, Z. Shengli, and C. Jun-Hong, "Prospects and problems of wireless communication for underwater sensor networks," *Wireless Communications and Mobile Computing*, vol. 8, no. 8, pp. 977–994, 2010.

[3] L. Li, M. C. Vuran, and I. F. Akyildiz, "Characteristics of underground channel for wireless underground sensor networks," presented at the *Sixth Annual Mediterranean Ad Hoc Networking WorkShop*, Corfu, Greece, June 12–15, 2007.

[4] J. J. Sojdehei, P. N. Wrathall, and D. F. Dinn, "Magneto-inductive (MI) communications," in *MTS/IEEE OCEANS, 2001. An Ocean Odyssey Conference and Exhibition*, vol. 1. pp. 513–519, 2001.

[5] S. Zhi, W. Pu, M. C. Vuran, M. A. Al-Rodhaan, A. M. Al-Dhelaan, and I. Akyildiz, "MISE-PIPE: Magnetic induction-based wireless sensor networks for underground pipeline monitoring," *Ad Hoc Networks*, vol. 9, no. 3, pp. 218–227, 2011.

[6] M. C. Domingo, "Magnetic induction for underwater wireless communication networks," *IEEE Transactions on Antennas and Propagation*, vol. 60, no. 6, pp. 2929–2939, 2012.

[7] B. Gulbahar and O. B. Akan, "A communication theoretical modeling and analysis of underwater magneto-inductive wireless channels," *IEEE Transactions on Wireless Communications*, vol. 11, no. 9, pp. 3326–3334, 2012.

[8] V. Parameswaran, H. Zhou, and Z. Zhang, "Irrigation control using wireless underground sensor networks," in *Sixth International Conference on Sensing Technology*, Kolkata, pp. 653–659, 2012.

[9] S. A. Meybodi, M. Dohler, A. N. Askarpour, J. Bendtsen, and J. D. Nielsen, "The feasibility of communication among pumps in a district heating system," *IEEE Antennas and Propagation Magazine*, vol. 55, no. 3, pp. 118–134, 2013.

[10] A. Markham and N. Trigoni, "Magneto-inductive networked rescue system (MINERS): Taking sensor networks underground," in *2012 ACM/IEEE Proceedings of the 11th International Conference on Information Processing in Sensor Networks*, Beijing, April 16–20, pp. 317–328, 2012.

[11] T. Abrudan, O. Kypris, N. Trigoni, and A. Markham, *Magneto-Inductive Underground Tracking: Principles and Systems*, ch. 8.2. Elsevier, 2016.

[12] T. E. Abrudan, Z. Xiao, A. Markham, and N. Trigoni, "Distortion rejecting magneto-inductive three-dimensional localization (MagLoc)," *IEEE Journal on Selected Areas in Communications*, vol. 33, no. 11, pp. 2404–2417, 2015.

[13] C. A. Balanis, *Antenna Theory*. Wiley, 2005.

[14] D. Halliday, R. Resnick, and J. Walker, *Fundamentals of Physics, Extended*, 3rd ed. Wiley, 1988.

[15] F. Sun, X. Ge, and S. He, "Can Maxwell's fish eye lens really give perfect imaging? Part II. The case with passive drains," *Progress in Electromagnetics Research*, vol. 110, pp. 313–328, 2010.

[16] Z. Sun, I. Akyildiz, S. Kisseleff, and W. Gerstacker, "Increasing the capacity of magnetic induction communications in RF-challenged environments," *IEEE Transactions on Communications*, vol. 61, no. 9, pp. 3943–3952, 2013.

[17] S. Kisseleff, W. Gerstacker, R. Schober, Z. Sun, and I. F. Akyildiz, "Channel capacity of magnetic induction based wireless underground sensor networks under practical constraints," in *2013 IEEE Wireless Communications and Networking Conference (WCNC)*, Shanghai, China, pp. 2603–2608, 2013.

[18] R. Gabillard, P. Degauque, and J. R. Wait, "Subsurface electromagnetic telecommunication: A review," *IEEE Transactions on Communication Technology*, vol. 19, no. 6, pp. 1217–1228, 1971.

[19] D. S. Jones, *The Theory of Electromagnetism*. Pergamon Press, 1964.

[20] H. Guo, Z. Sun, and P. Wang, "Channel modeling of MI underwater communication using tri-directional coil antenna," in *GLOBECOM: IEEE Global Communications Conference*, San Diego, CA, USA, pp. 1–6, 2015.

[21] Z. Sun and I. F. Akyildiz, "Magnetic induction communications for wireless underground sensor networks," *IEEE Transactions on Antennas and Propagation*, vol. 58, no. 7, pp. 2426–2435, 2010.

[22] M. Masihpour and J. I. Agbinya, "Cooperative relay in near field magnetic induction: A new technology for embedded medical communication systems," in *2010 Fifth International Conference on Broadband and Biomedical Communications*, Malaga, Spain, December 15–17, 2010.

[23] S. Kisseleff, B. Sackenreuter, I. F. Akyildiz, and W. Gerstacker, "On capacity of active relaying in magnetic induction based wireless underground sensor networks," in *2015 IEEE International Conference on Communications (ICC)*, London, UK, June 8–12, pp. 6541–6546, 2015.

[24] H. Ma, E. Liu, R. Wang, and X. Qu, "Effect of antenna deployment on achievable rate in cooperative magnetic induction communication," *IEEE Communications Letters*, vol. 23, no. 10, pp. 1748–1752, 2019.

[25] H. Ma, E. Liu, R. Wang, X. Yin, Z. Xu, X. Qu, and B. Li, "Antenna optimization for decode-and-forward relay in magnetic induction communications," *IEEE Transactions on Vehicular Technology*, vol. 69, no. 3, pp. 3449–3453, 2020.

[26] A. D. W. Gibson, "Channel characterisation and system design for sub-surface communications," PhD Thesis, University of Leeds, 2004.

[27] Z. Sun, I. F. Akyildiz, S. Kisseleff, and W. Gerstacker, "Increasing the capacity of magnetic induction communications in RF-challenged environments," *IEEE Transactions on Communications*, vol. 61, no. 9, pp. 3943–3952, 2013.

[28] H. Friis, "Noise figures of radio receivers," *Proceedings of the IRE*, vol. 32, no. 7, pp. 419–422, 1944.

[29] X. Tan and Z. Sun, "Environment-aware indoor localization using magnetic induction," in *GLOBECOM: 2015 IEEE Global Communications Conference*, San Diego, CA, USA, December 6–10, pp. 1–6, 2015.

[30] Z. Sun and I. F. Akyildiz, "Magnetic induction communications for wireless underground sensor networks," *IEEE Transactions on Antenna and Propagation*, vol. 58, no. 7, pp. 2426–2435, 2010.

[31] Z. Zhang, E. Liu, X. Qu, R. Wang, H. Ma, and Z. Sun, "Connectivity of magnetic induction-based ad hoc networks," *IEEE Transactions on Wireless Communications*, vol. 16, no. 7, pp. 4181–4191, 2017.

[32] H. Ma, E. Liu, R. Wang, and X. Qu, "Channel characteristics for vehicle magnetic induction communication with road disturbance," *IEEE Communications Letters*, vol. 24, no. 7, pp. 1363–1367, 2020.

[33] A. Karlsson, "Physical limitations of antennas in a lossy medium," *IEEE Transactions on Antennas and Propagation*, vol. 52, no. 8, pp. 2027–2033, 2004.

[34] C. A. Balanis, *Antenna Theory: Analysis and Design*. Wiley, 2005, vol. 1.

[35] Y. Rahmat-Samii, L. I. Williams, and R. G. Yaccarino, "The UCLA bi-polar planar-near-field antenna-measurement and diagnostics range," *IEEE Antennas and Propagation Magazine*, vol. 37, no. 6, pp. 16–35, 1995.

[36] C. Balanis, *Antenna Theory: Analysis and Design*. Wiley, 2015.

[37] C. Balanis, *Engineering Electromagnetics*. Wiley, 1989.

[38] U. Azad and Y. E. Wang, "Direct antenna modulation (DAM) for enhanced capacity performance of near-field communication (NFC) link," *IEEE Transactions on Circuits and Systems I: Regular Papers*, vol. 61, no. 3, pp. 902–910, 2014.

[39] S. Kisseleff, I. F. Akyildiz, and W. H. Gerstacker, "Digital signal transmission in magnetic induction based wireless underground sensor networks," *IEEE Transactions on Communications*, vol. 63, no. 6, pp. 2300–2311, 2015.

[40] H. Guo, Z. Sun, and P. Wang, "Multiple frequency band channel modeling and analysis for magnetic induction communication in practical underwater environments," *IEEE Transactions on Vehicular Technology*, vol. 66, no. 8, pp. 6619–6632, 2017.

[41] H. Guo and Z. Sun, "Channel and energy modeling for self-contained wireless sensor networks in oil reservoirs," *IEEE Transactions on Wireless Communications*, vol. 13, no. 4, pp. 2258–2269, 2014.

[42] J. Jadidian and D. Katabi, "Magnetic MIMO: How to charge your phone in your pocket," in *Proceedings of the 20th Annual International Conference on Mobile Computing and Networking*, pp. 495–506, 2014.

[43] G. Yang, M. R. V. Moghadam, and R. Zhang, "Magnetic MIMO signal processing and optimization for wireless power transfer," *IEEE Transactions on Signal Processing*, vol. 65, no. 11, pp. 2860–2874, 2017.

[44] H.-D. Lang, A. Ludwig, and C. D. Sarris, "Convex optimization of wireless power transfer systems with multiple transmitters," *IEEE Transactions on Antennas and Propagation*, vol. 62, no. 9, pp. 4623–4636, 2014.

[45] P. Grover and A. Sahai, "Shannon meets Tesla: Wireless information and power transfer," in *2010 IEEE International Symposium on Information Theory Proceedings (ISIT)*, pp. 2363–2367, 2010.

[46] H. Guo and Z. Sun, "Increasing the capacity of magnetic induction communication using MIMO coil-array," in *2016 IEEE Global Communications Conference (GLOBECOM)*, pp. 1–6, 2016.

[47] R. W. Ziolkowski and A. Erentok, "Metamaterial-based efficient electrically small antennas," *IEEE Transactions on Antennas and Propagation*, vol. 54, no. 7, pp. 2113–2130, 2006.

[48] D. R. Smith, J. B. Pendry, and M. C. Wiltshire, "Metamaterials and negative refractive index," *Science*, vol. 305, no. 5685, pp. 788–792, 2004.

[49] E. Shamonina, V. Kalinin, K. Ringhofer, and L. Solymar, "Magnetoinductive waves in one, two, and three dimensions," *Journal of Applied Physics*, vol. 92, no. 10, pp. 6252–6261, 2002.

[50] L. Solymar and E. Shamonina, *Waves in Metamaterials*. Oxford University Press, 2009.

[51] H. Guo, Z. Sun, J. Sun, and N. M. Litchinitser, "M^2I: Channel modeling for metamaterial-enhanced magnetic induction communications," *IEEE Transactions on Antennas and Propagation*, vol. 63, no. 11, pp. 5072–5087, 2015.

[52] V. G. Veselago, "The electrodynamics of substances with simultaneously negative values of and μ," *Soviet Physics Uspekhi*, vol. 10, no. 4, p. 509, 1968.

[53] J. B. Pendry, "Negative refraction makes a perfect lens," *Physical Review Letters*, vol. 85, no. 18, p. 3966, 2000.

[54] Y. Urzhumov and D. R. Smith, "Metamaterial-enhanced coupling between magnetic dipoles for efficient wireless power transfer," *Physical Review B*, vol. 83, no. 20, p. 205114, 2011.

[55] R. W. Ziolkowski and A. D. Kipple, "Application of double negative materials to increase the power radiated by electrically small antennas," *IEEE Transactions on Antennas and Propagation*, vol. 51, no. 10, pp. 2626–2640, 2003.

[56] S. Arslanagic, R. W. Ziolkowski, and O. Breinbjerg, "Analytical and numerical investigation of the radiation from concentric metamaterial spheres excited by an electric Hertzian dipole," *Radio Science*, vol. 42, no. 06, pp. 1–20, 2007.

[57] N. Engheta and R. W. Ziolkowski, *Metamaterials: Physics and Engineering Explorations*. Wiley, 2006.

[58] H. Guo, Z. Sun, and C. Zhou, "Practical design and implementation of metamaterial-enhanced magnetic induction communication," *IEEE Access*, vol. 5, pp. 17213–17229, 2017.

[59] T. C. Choy, *Effective Medium Theory: Principles and Applications*, vol. 165. Oxford University Press, 2015.

[60] J. B. Pendry, A. J. Holden, D. Robbins, and W. Stewart, "Magnetism from conductors and enhanced nonlinear phenomena," *IEEE Transactions on Microwave Theory and Techniques*, vol. 47, no. 11, pp. 2075–2084, 1999.

[61] C. Scarborough, Z. Jiang, D. Werner, C. Rivero-Baleine, and C. Drake, "Experimental demonstration of an isotropic metamaterial super lens with negative unity permeability at 8.5 MHz," *Applied Physics Letters*, vol. 101, no. 1, p. 014101, 2012.

[62] G. Lipworth, J. Ensworth, K. Seetharam, D. Huang, J. S. Lee, P. Schmalenberg, T. Nomura, M. S. Reynolds, D. R. Smith, and Y. Urzhumov, "Magnetic metamaterial superlens for increased range wireless power transfer," *Scientific Reports*, vol. 4, no. 1, pp. 1–6, 2014.

[63] R. W. Ziolkowski and A. D. Kipple, "Application of double negative materials to increase the power radiated by electrically small antennas," *IEEE Transactions on Antenna and Propagation*, vol. 51, no. 10, pp. 2626–2640, 2003.

[64] R. W. Ziolkowski and A. Erentok, "At and below the chu limit: Passive and active broad bandwidth metamaterial-based electrically small antennas," *IET Microwave, Antennas and Propagation*, vol. 1, no. 1, pp. 116–128, 2007.

[65] M. Rubinov and O. Sporns, "Complex network measures of brain connectivity: Uses and interpretations," *NeuroImage*, vol. 52, no. 3, pp. 1059–1069, 2010, Computational Models of the Brain. [Online]. Available: www.sciencedirect.com/science/article/pii/S105381190901074X

[66] N. A. M. Araújo and H. J. Herrmann, "Explosive percolation via control of the largest cluster," *Physical Review Letters*, vol. 105, p. 035701, 2010. [Online]. Available: https://link.aps.org/doi/10.1103/PhysRevLett.105.035701

[67] D. Urban and T. Keitt, "Landscape connectivity: A graph-theoretic perspective," *Ecology*, vol. 82, no. 5, pp. 1205–1218, 2001. [Online]. Available: https://esajournals.onlinelibrary.wiley.com/doi/abs/10.1890/0012-9658%282001%29082%5B1205%3ALCAGTP%5D2.0.CO%3B2

[68] Y. Shang, W. Ruml, Y. Zhang, and M. Fromherz, "Localization from connectivity in sensor networks," *IEEE Transactions on Parallel and Distributed Systems*, vol. 15, no. 11, pp. 961–974, 2004.

[69] P. Gupta and P. R. Kumar, "Critical power for asymptotic connectivity," in *Proceedings of the IEEE CDC*, Tampa, USA, December 1998.

[70] C. Bettstetter, "On the minimum node degree and connectivity of a wireless multihop network," in *Proceedings of the ACM MOBIHOC '02*, Lausanne, Switzerland, June 2002.

[71] R. Wattenhofer, L. Li, P. Bahl, and Y.-M. Wang, "Distributed topology control for power efficient operation in multihop wireless ad hoc networks," in *IEEE INFOCOM*, vol. 3, Anchorage, AK, pp. 1388–1397, 2001.

[72] X. Wang, X. Lin, Q. Wang, and W. Luan, "Mobility increases the connectivity of wireless networks," *IEEE/ACM Transactions on Networking*, vol. 21, no. 2, pp. 440–454, 2013.

[73] C. Bettstetter, C. Hartmann, and C. Moser, "How does randomized beamforming improve the connectivity of ad hoc networks?" in *2005 IEEE International Conference on Communications, 2005. ICC 2005*, vol. 5. pp. 3380–3385, 2005.

[74] X. Zhou, S. Durrani, and H. M. Jones, "Connectivity of ad hoc networks: Is fading good or bad?" in *2nd International Conference on Signal Processing and Communication Systems*, Gold Coast, pp. 1–5, 2008.

[75] L. Wang, P.-J. Wan, and W. Washington, "Connectivity of multihop wireless networks with log-normal shadowing," *Wireless Networks*, vol. 21, no. 7, pp. 2279–2292, 2015.

[76] C. Bettstetter, C. Hartmann, and C. Moser, "How does randomized beamforming improve the connectivity of ad hoc networks?" in *IEEE ICC*, vol. 5, Corunna, Spain, pp. 3380–3385, 2005.

[77] L. T. Dung and B. An, "A modeling framework for supporting and evaluating connectivity in cognitive radio ad hoc networks with beamforming," *Wireless Networks*, vol. 23, pp. 1743–1755, 2017.

[78] P. N. Wrathall and J. J. Sojdehei, "Magneto-inductive communications," in *AeroSense*, pp. 229–236, 1999.

[79] V. Rožanskij and V. B. Rojansky, *Electromagnetic Fields and Waves*. Dover Publications, 1971.

[80] Z. Zhang, E. Liu, X. Qu, D. Liu, R. Wang, and F. Liu, "Effective coverage for the connectivity of magnetic induction-based ad hoc networks," in *IEEE GLOBECOM*, San Diego, pp. 1–6, 2015.

[81] Z. Sun and I. F. Akyildiz, "Wireless underground networks," *Foundations and Trends in Networking*, vol. 5, no. 4, pp. 138–283, 2012.

[82] S. Kisseleff, W. Gerstacker, Z. Sun, and I. Akyildiz, "On the throughput of wireless underground sensor networks using magneto-inductive waveguides," in *2013 IEEE Global Communications Conference (GLOBECOM)*, pp. 322–328, December 2013.

[83] C. Bettstetter, "On the connectivity of ad hoc networks," *The Computer Journal*, vol. 47, no. 4, pp. 432–447, 2004.

[84] C. Bettstetter, "On the minimum node degree and connectivity of a wireless multihop network," in *ACM MOBIHOC*, Lausanne, Switzerland, pp. 80–91, 2002.

[85] S. Boyd and L. Vandenberghe, *Convex Optimization*. Cambridge University Press, 2004.

[86] D. Barry, J.-Y. Parlange, L. Li, H. Prommer, C. Cunningham, and F. Stagnitti, "Analytical approximations for real values of the Lambert W-function," *Mathematics and Computers in Simulation*, vol. 53, no. 1, pp. 95–103, 2000.

[87] S. Kisseleff, B. Sackenreuter, I. F. Akyildiz, and W. Gerstacker, "On capacity of active relaying in magnetic induction based wireless underground sensor networks," in *Proceedings of the IEEE International Conference on Communications*, London, UK, pp. 6541–6546, June 2015.

[88] I. F. Akyildiz, Z. Sun, and M. C. Vuran, "Signal propagation techniques for wireless underground communication networks," *Physical Communication*, vol. 2, no. 3, pp. 167–183, 2009. [Online]. Available: www.sciencedirect.com/science/article/pii/S1874490709000408

[89] Z. Sun, P. Wang, M. C. Vuran, M. A. Al-Rodhaan, A. M. Al-Dhelaan, and I. F. Akyildiz, "BorderSense: Border patrol through advanced wireless sensor networks," *Ad Hoc Networks*, vol. 9, no. 3, pp. 468–477, 2011. [Online]. Available: www.sciencedirect.com/science/article/pii/S1570870510001484

[90] S. Zhi and I. F. Akyildiz, "Magnetic induction communications for wireless underground sensor networks," *IEEE Transactions on Antennas & Propagation*, vol. 58, no. 7, pp. 2426–2435, 2010.

[91] Z. Sun and I. F. Akyildiz, "Underground wireless communication using magnetic induction," in *IEEE International Conference on Communications*, 2009.

[92] Z. Sun and I. F. Akyildiz, "Optimal deployment for magnetic induction-based wireless networks in challenged environments," *IEEE Transactions on Wireless Communications*, vol. 12, no. 3, pp. 996–1005, 2013.

[93] O. Schwarzkopf, "Computational geometry: Algorithms and applications," *Library Philosophy & Practice*, vol. 2011, no. 1, pp. 333–334, 2000.

[94] R. Bansal, "Near-field magnetic communication," *IEEE Antennas & Propagation Magazine*, vol. 46, no. 2, pp. 114–115, 2004.

[95] R. R. A. Syms, I. R. Young, and L. Solymar, "Low-loss magneto-inductive waveguides," *Journal of Physics D Applied Physics*, vol. 39, no. 18, pp. 3945–3951, 2006.

[96] E. Shamonina, V. A. Kalinin, K. H. Ringhofer, and L. Solymar, "Magnetoinductive waves in one, two, and three dimensions," *Journal of Applied Physics*, vol. 92, no. 10, p. 6252, 2002.

[97] O. Sydoruk, O. Zhuromskyy, and E. Shamonina, "Dispersion characteristics of magneto-inductive waves made up by doubly periodic elements," in *Conference on Metamaterials; 20050830-31*; Warsaw, PL, 2005.

[98] R. Ramanathan and R. Rosales-Hain, "Topology control of multiple wireless networks using transmit power adjustment," in *IEEE Proceedings. INFOCOM 2000. Nineteenth Annual Joint Conference of the IEEE Computer and Communications Societies*, 2000.

[99] M. Burkhart, P. Von Rickenbach, R. Wattenhofer, and A. Zollinger, "Does topology control reduce interference?" *Department of Computer Science ETH Zurich 8092 Zurich*, Switzerland, 2004.

[100] J. Pan, Y. T. Hou, L. Cai, Y. Shi, and S. X. Shen, "Topology control for wireless sensor networks," in *Proceedings of the 9th Annual International Conference on Mobile Computing and Networking*, ser. MobiCom '03. New York, NY,

USA: Association for Computing Machinery, pp. 286–299, 2003. [Online]. Available: https://doi.org/10.1145/938985.939015

[101] Y. Wang, "Topology control for wireless sensor networks," in *Wireless Sensor Networks and Applications. Signals and Communication Technology*. Springer, 2008.

[102] A. Khan, "Sensor deployment and target localization based on virtual forces," in *IEEE INFOCOM 2003. Twenty-Second Annual Joint Conference of the IEEE Computer and Communications Societies*, 2014.

[103] G. Wang, G. Cao, and T. La Porta, "Movement-assisted sensor deployment," *IEEE Transactions on Mobile Computing*, vol. 5, no. 6, pp. 640–652, 2006.

[104] Z. Sun and I. F. Akyildiz, "On capacity of magnetic induction-based wireless underground sensor networks," in *2012 Proceedings IEEE INFOCOM*, pp. 370–378, 2012.

[105] C. P. Dettmann and O. Georgiou, "Product of n independent uniform random variables," *Statistics & Probability Letters*, vol. 79, no. 24, pp. 2501–2503, 2009.

[106] B. Y. Wu and K. M. Chao, *Spanning Trees and Optimization Problems*. Spanning Trees and Optimization Problems, 2004.

[107] D. B. Mark, *Computational Geometry: Algorithms and Applications,* 3rd ed. Springer, 2013.

[108] I. F. Akyildiz and E. P. Stuntebeck, "Wireless underground sensor networks: Research challenges," *Ad Hoc Networks*, vol. 4, no. 6, pp. 669–686, 2006.

[109] S. Vuran, "Signal propagation techniques for wireless underground communication networks," *Physical Communication*, vol. 2, no. 3, pp. 167–183 2009.

[110] A. Karalis, J. D. Joannopoulos, and M. Soljačić, "Efficient wireless non-radiative mid-range energy transfer," *Annals of Physics*, vol. 323, no. 1, pp. 34–48, 2008.

[111] Z. Sun and I. F. Akyildiz, "On capacity of magnetic induction-based wireless underground sensor networks," in *2012 Proceedings IEEE INFOCOM*, pp. 370–378, 2012.

[112] A. R. Silva and M. C. Vuran, "Development of a testbed for wireless underground sensor networks," *EURASIP Journal on Wireless Communications and Networking*, vol. 2010, pp. 1–14, 2010.

[113] S. Zhi and I. F. Akyildiz, "Connectivity in wireless underground sensor networks," in *Proceedings of the Seventh Annual IEEE Communications Society Conference on Sensor, Mesh and Ad Hoc Communications and Networks, SECON 2010*, Boston, MA, USA, June 21–25, 2010.

[114] M. Wiltshire, E. Shamonina, I. R. Young, and L. Solymar, "Dispersion characteristics of magneto-inductive waves: Comparison between theory and experiment," *Electronics Letters*, vol. 39, no. 2, pp. 215–217, 2003.

[115] S. Zhi and I. F. Akyildiz, "Deployment algorithms for wireless underground sensor networks using magnetic induction," in *Global Telecommunications Conference*, 2010.

[116] P. Gupta and P. R. Kumar, "The capacity of wireless networks," *IEEE Transactions on Information Theory*, vol. 46, no. 2, pp. 388–404, 2000.

[117] C. X. Wang, X. Hong, H. H. Chen, and J. Thompson, "On capacity of cognitive radio networks with average interference power constraints," *IEEE Transactions on Wireless Communications*, vol. 8, no. 4, pp. 1620–1625, 2009.

[118] V. Bhandari and N. H. Vaidya, "Capacity of multi-channel wireless networks with random (c,f) assignment," *Technical Report*, 2007.

[119] C. Jiang, Y. Shi, Y. T. Hou, and S. Kompella, "On the asymptotic capacity of multi-hop MIMO ad hoc networks," *IEEE Transactions on Wireless Communications*, vol. 10, no. 4, pp. 1032–1037, 2011.

[120] A. Spyropoulos and C. S. Raghavendra, "Capacity bounds for ad-hoc networks using directional antennas," in *IEEE International Conference on Communications*, 2003.

[121] Y. Yemini, "The OSI network management model," *IEEE Communications Surveys & Tutorials*, vol. 3, no. 1, pp. 20–29, 1993.

[122] B. A. A. Nunes, M. Mendonca, X.-N. Nguyen, K. Obraczka, and T. Turletti, "A survey of software-defined networking: Past, present, and future of programmable networks," *IEEE Communications Surveys & Tutorials*, vol. 16, no. 3, pp. 1617–1634, 2014.

[123] F. Tang, Z. M. Fadlullah, B. Mao, and N. Kato, "An intelligent traffic load prediction-based adaptive channel assignment algorithm in SDN-IoT: A deep learning approach," *IEEE Internet of Things Journal*, vol. 5, no. 6, pp. 5141–5154, 2018.

[124] Z. Qin, G. Denker, C. Giannelli, P. Bellavista, and N. Venkatasubramanian, "A software defined networking architecture for the internet-of-things," in *2014 IEEE network operations and management symposium (NOMS)*, pp. 1–9, 2014.

[125] H. Guo and Z. Sun, "Demo abstract: Prototyping M^2I communication system for underground and underwater networks," in *2017 IEEE Conference on Computer Communications Workshops (INFOCOM WKSHPS)*, pp. 962–963, 2017.

[126] R. Vyas and B. Tye, "A sequential RFID system for robust communication with underground carbon steel pipes in oil and gas applications," *Electronics*, vol. 8, no. 12, p. 1374, 2019.

[127] NFC Forum, "NFC forum technical specification," 2006.

[128] D. K. Klair, K.-W. Chin, and R. Raad, "A survey and tutorial of RFID anti-collision protocols," *IEEE Communications Surveys & Tutorials*, vol. 12, no. 3, pp. 400–421, 2010.

[129] D. Delen, R. Sharda, and B. C. Hardgrave, "The promise of RFID-based sensors in the perishables supply chain," *IEEE Wireless Communications*, vol. 18, no. 2, pp. 82–88, 2011.

[130] H. Liu, M. Bolic, A. Nayak, and I. Stojmenovi, "Integration of RFID and wireless sensor networks," in *Encyclopedia on Ad Hoc and Ubiquitous Computing: Theory and Design of Wireless Ad Hoc, Sensor, and Mesh Networks*. World Scientific, pp. 319–347, 2010.

[131] A. Nosratinia, T. E. Hunter, and A. Hedayat, "Cooperative communication in wireless networks," *IEEE Communications Magazine*, vol. 42, no. 10, pp. 74–80, 2004.

[132] T. Riihonen, S. Werner, and R. Wichman, "Mitigation of loopback self-interference in full-duplex MIMO relays," *IEEE Transactions on Signal Processing*, vol. 59, no. 12, pp. 5983–5993, 2011.

[133] I. Krikidis, H. A. Suraweera, P. J. Smith, and C. Yuen, "Full-duplex relay selection for amplify-and-forward cooperative networks," *IEEE Transactions on Wireless Communications*, vol. 11, no. 12, pp. 4381–4393, 2012.

[134] E. Shamonina, V. Kalinin, K. Ringhofer, and L. Solymar, "Magneto-inductive waveguide," *Electronics Letters*, vol. 38, no. 8, pp. 371–373, 2002.

[135] H. Guo and Z. Sun, "Full-duplex metamaterial-enabled magnetic induction networks in extreme environments," in *IEEE INFOCOM 2018-IEEE Conference on Computer Communications*, pp. 558–566, 2018.

[136] M. J. Neely and E. Modiano, "Capacity and delay tradeoffs for ad hoc mobile networks," *IEEE Transactions on Information Theory*, vol. 51, no. 6, pp. 1917–1937, 2005.

[137] K. Kant and A. Pal, "Internet of perishable logistics," *IEEE Internet Computing*, vol. 21, no. 1, pp. 22–31, 2017.

[138] P. Gupta and P. R. Kumar, "The capacity of wireless networks," *IEEE Transactions on Information Theory*, vol. 46, no. 2, pp. 388–404, 2000.

[139] M. Franceschetti, O. Dousse, N. David, and P. Thiran, "Closing the gap in the capacity of wireless networks via percolation theory," *IEEE Transactions on Information Theory*, vol. 53, no. 3, pp. 1009–1018, 2007.

[140] S. Toumpis and A. J. Goldsmith, "Large wireless networks under fading, mobility, and delay constraints," in *INFOCOM 2004. Twenty-third Annual Joint Conference of the IEEE Computer and Communications Societies*, vol. 1, 2004.

[141] F. H. Raab, E. B. Blood, T. O. Steiner, and H. R. Jones, "Magnetic position and orientation tracking system," *IEEE Transactions on Aerospace and Electronic Systems*, vol. AES-15, no. 5, pp. 709–718, 1979.

[142] E. Paperno, I. Sasada, and E. Leonovich, "A new method for magnetic position and orientation tracking," *IEEE Transactions on Magnetics*, vol. 37, no. 4, pp. 1938–1940, 2001.

[143] E. Paperno and P. Keisar, "Three-dimensional magnetic tracking of biaxial sensors," *IEEE Transactions on Magnetics*, vol. 40, no. 3, pp. 1530–1536, 2004.

[144] A. Sheinker, B. Ginzburg, N. Salomonski, L. Frumkis, and B. Kaplan, "Localization in 2D using beacons of low frequency magnetic field," *IEEE Journal of Selected Topics in Applied Earth Observations and Remote Sensing*, vol. 6, no. 2, pp. 1020–1030, 2013.

[145] A. Sheinker, B. Ginzburg, N. Salomonski, L. Frumkis, and B. Kaplan, "Localization in 3-D using beacons of low frequency magnetic field," *IEEE Transactions on Instrumentation and Measurement*, vol. 62, no. 12, pp. 3194–3201, 2013.

[146] A. Markham, N. Trigoni, D. W. Macdonald, and S. A. Ellwood, "Underground localization in 3-D using magneto-inductive tracking," *IEEE Sensors Journal*, vol. 12, no. 6, pp. 1809–1816, 2012.

[147] M. Q.-H. Meng, T. Mei, J. Pu, C. Hu, X. Wang, and Y. Chan, "Wireless robotic capsule endoscopy: State-of-the-art and challenges," in *Proceedings of the Fifth World Congress Intelligent Control and Automation (IEEE Cat. No.04EX788)*, vol. 6, pp. 5561–555a, 2004.

[148] C. Hu, M. Q.-H. Meng, M. Mandal, and X. Wang, "3-Axis magnetic sensor array system for tracking magnet's position and orientation," in *Proceedings of the 6th World Congress Intelligent Control and Automation*, vol. 2, pp. 5304–5308, 2006.

[149] M. Li, C. Hu, S. Song, H. Dai, and M. Q.-H. Meng, "Detection of weak magnetic signal for magnetic localization and orientation in capsule endoscope," in *Proceedings of the IEEE International Conference on Automation and Logistics*, pp. 900–905, 2009.

[150] S. Yabukami, H. Kikuchi, M. Yamaguchi, K. I. Arai, K. Takahashi, A. Itagaki, and N. Wako, "Motion capture system of magnetic markers using three-axial magnetic field sensor," *IEEE Transactions on Magnetics*, vol. 36, no. 5, pp. 3646–3648, 2000.

[151] X. Tan, Z. Sun, and P. Wang, "On localization for magnetic induction-based wireless sensor networks in pipeline environments," in *Proceedings of the IEEE International Conference on Communications (ICC)*, pp. 2780–2785, 2015.

[152] S. Kisseleff, I. F. Akyildiz, and W. H. Gerstacker, "Survey on advances in magnetic induction-based wireless underground sensor networks," *IEEE Internet of Things Journal*, vol. 5, no. 6, pp. 4843–4856, 2018.

[153] Y. Li, S. Wang, C. Jin, Y. Zhang, and T. Jiang, "A survey of underwater magnetic induction communications: Fundamental issues, recent advances, and challenges," *IEEE Communications Surveys Tutorials*, vol. 21, no. 3, pp. 2466–2487, 2019.

[154] N. Saeed, M.-S. Alouini, and T. Y. Al-Naffouri, "Toward the internet of underground things: A systematic survey," *IEEE Communications Surveys Tutorials*, vol. 21, no. 4, pp. 3443–3466, 2019.

[155] R. A. Khalil, N. Saeed, M. I. Babar, and T. Jan, "Toward the internet of underwater things: Recent developments and future challenges," *IEEE Consumer Electronics Magazine*, vol. 10, no. 6, pp. 32–37, 2020.

[156] I. F. Akyildiz and E. P. Stuntebeck, "Akyildiz wireless underground sensor networks: Research challenges," *Ad Hoc Networks*, vol. 4, no. 6, pp. 669–686, 2006.

[157] Z. Zhang, E. Liu, X. Zheng, Y. Jian, D. Wang, and D. Liu, "Cooperative magnetic induction based through-the-earth communication," in *IEEE/CIC ICCC*, pp. 653–657, 2014.

[158] Z. Sun, I. F. Akyildiz, and P. Wang, "Realizing underwater communication through magnetic induction," *IEEE Communications Magazine: Articles, News, and Events of Interest to Communications Engineers*, 2015.

[159] R. A. Khalil and N. Saeed, "Optimal relay placement in magnetic induction-based internet of underwater things," *IEEE Sensors Journal*, vol. 21, no. 1, pp. 821–828, 2020.

[160] D. Wei, L. Yan, C. Huang, J. Wang, J. Chen, M. Pan, and Y. Fang, "Dynamic magnetic induction wireless communications for autonomous-underwater-vehicle-assisted underwater IoT," *IEEE Internet of Things Journal*, vol. 7, no. 10, pp. 9834–9845, 2020.

[161] T. AL-Kadi, Z. AL-Tuwaijri, and A. AL-Omran, "Wireless sensor networks for leakage detection in underground pipelines: A survey paper," *Procedia Computer Science*, vol. 21, pp. 491–498, 2013.

[162] Z. Sun, P. Wang, M. C. Vuran, M. A. Al-Rodhaan, A. M. Al–Dhelaan, and I. F. Akyildiz, "Mise-pipe: Magnetic induction-based wireless sensor networks for underground pipeline monitoring," *Ad Hoc Networks*, vol. 9, no. 3, pp. 218–227, 2011.

[163] A. Markham and N. Trigoni, "Proceedings of the 11th international conference on information processing in sensor networks: 11th international conference on information processing in sensor networks," in *Magneto-Inductive NEtworked Rescue System (MINERS): Taking Sensor Networks Underground*, Beijing, China, April 2012.

[164] M. A. Gutierrez-Estevez, U. Krüger, K. A. Krueger, K. Manolakis, V. Jungnickel, K. Jaksch, K. Krueger, S. Mikulla, R. Giese, M. Sohmer et al., "Acoustic broadband communications over deep drill strings using adaptive OFDM," in *2013 IEEE Wireless Communications and Networking Conference (WCNC)*, pp. 4089–4094, 2013.

[165] C. H. Martins, A. A. Alshehri, and I. F. Akyildiz, "Novel mi-based (FracBot) sensor hardware design for monitoring hydraulic fractures and oil reservoirs," in *2017 IEEE 8th Annual Ubiquitous Computing, Electronics and Mobile Communication Conference (UEMCON)*, pp. 434–441, 2017.

[166] A. A. Al-Shehri, I. F. Akyildiz, J. M. Servin, and H. K. Schmidt, "Fracbot technology for mapping hydraulic fractures," in *SPE Annual Technical Conference and Exhibition*, OnePetro, 2017.

[167] T. H. Lee, *The Design of CMOS Radio-Frequency Integrated Circuits*. Cambridge University Press, 2003.

[168] M. Zargham and P. G. Gulak, "Maximum achievable efficiency in near-field coupled power-transfer systems," *IEEE Transactions on Biomedical Circuits and Systems*, vol. 6, no. 3, pp. 228–245, 2012.

[169] H.-D. Lang and C. D. Sarris, "Optimization of wireless power transfer systems enhanced by passive elements and metasurfaces," *IEEE Transactions on Antennas and Propagation*, vol. 65, no. 10, pp. 5462–5474, 2017.

[170] Z. Li and Z. Sun, "Enabling magnetic beamforming in MIMO wireless power transfer using reconfigurable metasurface," in *GLOBECOM 2020–2020 IEEE Global Communications Conference*, pp. 1–6, 2020.

[171] Z. Li and Z. Sun, "Antenna system optimization for active metamaterial-enhanced magnetic induction communications," in *2019 13th European Conference on Antennas and Propagation (EuCAP)*, pp. 1–5, 2019.

[172] F. Gao, F. Zhang, M. Huang, and D. F. Sievenpiper, "Programmable screen for patterning magnetic fields," *IEEE Transactions on Microwave Theory and Techniques*, vol. 62, no. 3, pp. 481–490, 2014.

[173] Z. Li and Z. Sun, "Antenna system optimization for active metamaterial-enhanced magnetic induction communications," in *2019 13th European Conference on Antennas and Propagation (EuCAP)*, pp. 1–5, 2019.

[174] M. Haenggi, J. G. Andrews, F. Baccelli, O. Dousse and M. Franceschetti, "Stochastic geometry and random graphs for the analysis and design of wireless networks," *IEEE Journal on Selected Areas in Communications*, vol. 27, no. 7, pp. 1029–1046, 2009. [Online]. Available: https://doi.org/10.1109/JSAC.2009.090902

[175] L. Li, M. Vuran, and I. Akyildiz, "Characteristics of Underground Channel for Wireless Underground Sensor Networks," 2012.

Notation

j	$\sqrt{-1}$	a_0	lattice constant
k	propagation constant	$c_q(\ell)$	q-th order correlation function
θ	polar angle in Spherical Coordinates	d_{E}	embedding dimension
		d_f	fractal dimension
d	distance	L	system size
a	coil radius	$\equiv$	*defined* to be equal
I_0	coil current	$\sim$	*asymptotically* equal (in scaling sense)
AFM	atomic force microscope	$\approx$	*approximately* equal (in numerical value)
AKPZ	anisotropic KPZ equation		

Index

For EU product safety concerns, contact us at Calle de José Abascal, 56–1°,
28003 Madrid, Spain or eugpsr@cambridge.org.

www.ingramcontent.com/pod-product-compliance
Lightning Source LLC
LaVergne TN
LVHW062209200926
848484LV00006B/1363
* 9 7 8 1 1 0 8 4 8 1 6 7 0 *